Sustainability Sciences in Asia and Africa

Sustainable Agriculture and Food Security

This book series support the global efforts towards sustainability by providing timely coverage of the progress, opportunities, and challenges of sustainable food production and consumption in Asia and Africa. The series narrates the success stories and research endeavors from the regions of Africa and Asia on issues relating to SDG 2: Zero hunger. It fosters the research in transdisciplinary academic fields spanning across sustainable agriculture systems and practices, post- harvest and food supply chains. It will also focus on breeding programs for resilient crops, efficiency in crop cycle, various factors of food security, as well as improving nutrition and curbing hunger and malnutrition. The focus of the series is to provide a comprehensive publication platform and act as a knowledge engine in the growth of sustainability sciences with a special focus on developing nations. The series will publish mainly edited volumes but some authored volumes. These volumes will have chapters from eminent personalities in their area of research from different parts of the world.

Sugandha Munshi • Madhulika Singh
Editors

Women Farmers: Unheard Being Heard

Springer

Editors
Sugandha Munshi
Sustainable Impact Platform
International Rice Research Institute
Patna, Bihar, India

Madhulika Singh
Sustainable Agri-food Systems
International Maize and Wheat Improvement Center
Patna, Bihar, India

ISSN 2730-6771 ISSN 2730-678X (electronic)
Sustainability Sciences in Asia and Africa
ISSN 2730-6798 ISSN 2730-6801 (electronic)
Sustainable Agriculture and Food Security
ISBN 978-981-19-6977-5 ISBN 978-981-19-6978-2 (eBook)
https://doi.org/10.1007/978-981-19-6978-2

This Springer imprint is published by the registered company Springer Nature Singapore Pte Ltd.
The registered company address is: 152 Beach Road, #21-01/04 Gateway East, Singapore 189721, Singapore

Series Preface

The global food production systems require expanding their capacity to produce almost twice the current levels to safeguard the food security of the burgeoning population worldwide. More than 800 million suffering from undernourishment worldwide pose a great risk to the attainment of sustainable development goal (SDG) 2 of the UN that targets "End hunger, achieve food security and improved nutrition and promote sustainable agriculture" within the next seven years. The challenge is further exacerbated by the rising weather extremities and unpredictability in rainfall patterns and pest-pathogen dynamics associated with global climate change that has profound negative impact on the agricultural productivity and farm incomes worldwide. Also, the future targets of food production should be secured in a resource-constrained agricultural settings and with least environment footprint, thus calling for sustainable innovations in Agri-farming systems and enhanced participation of women in agriculture. The challenge to reduce hunger is alarming in the case of developing nations particularly Asia and Africa that house the largest proportion of people suffering from malnutrition and other nutrition-related issues. Furthermore, the agri-food systems in Asia and Africa are severely constrained by subsistence nature of farming, declining land and other agricultural resources, increasing environmental pollution, soil and biodiversity degradation, and climate change. Therefore, this book series, "Sustainable Agriculture and Food Security" has been planned to support the global efforts towards sustainability by providing timely coverage of the progress, opportunities, and challenges of sustainable food production and consumption in Asia and Africa. The series narrates the success stories and research endeavors from the regions of Africa and Asia on issues relating to SDG 2: Zero hunger. It fosters research in transdisciplinary academic fields spanning across sustainable agriculture systems and practices, post-harvest and food supply chains. The focus of the series is to provide a comprehensive publication platform and act as a knowledge engine in the growth of sustainability sciences with a special focus on developing nations.

In terms of agriculture in developing countries, about 80% of the world's food is produced by smallholding farmers. Women make up on average 43% of this agricultural labor in developing countries. While, in South Asia, more than two thirds of employed women work in agriculture, over half of farmers are women in Eastern Africa. Limited exposure and knowledge of resource-poor farmers, particularly the women farmers, restrict them to innovate their farming systems and practices in the face of climate change. Despite receiving marked discrimination and having less access to agricultural inputs and farm implements, banking, and financial services, women remain crucial to the smallholder farming systems, which have been reported to be 2–4% more effective in reducing hunger and poverty than other sectors. Furthermore, the women engaged in small-scale agriculture are more vulnerable to the risks posed by climate change. The food production by the women could go up by 30% if they have better access to resources and education, and overcoming the productivity barriers of hundreds of millions of women farmers could potentially eliminate the hunger-related issues of about 150 million people.

The women should be made aware of the evolving agricultural education and knowledge at an early age as one of the most promising solutions to the existing challenges that restrict their potential for reducing poverty and hunger. The women farmers should be trained with advanced farming tools and techniques by bringing the extension programs at the grassroots level and organization of government-sponsored training programs that are timely and affordable. Training programs should also focus on addressing the specific needs of the women farmers. *Greater engagement of trained smallholder women* farmer in agriculture backed by government policies will help developing reasonable solutions to the risks and challenges associated with climate change with respect to the resilience and productivity of farming systems. Developing *required* skills or capabilities among *women* and removing gender-based barriers is imperative to innovate and transform the agriculture sector in developing countries.

In view of this, the present book, *Women Farmers: Unheard Being Heard* edited by Sugandha Munshi and Madhulika Singh, puts a sincere effort to pen down the enormous contributions of women farmers to agriculture and food production in Asia. The book provides an opportunity to recognize and celebrate women farmer's role that remains overlooked or more practically "invisible" due to the strong patriarchal ecosystem refraining them to access land, improved *inputs, equipment, training, finance, and supportive resources*. One of the major constraints to creating an equitable agriculture system is land rights, and the present book highlights how changes in land rights of women in India since 2005 have contributed towards generating sustainable livelihood, improving agricultural productivity, and resilience of farming systems. The book shares evidence-based studies on the creation and adoption of innovative platforms such as self-help groups that have played a stronger role in empowering women in agriculture. The different chapters cover key challenges that women face in agriculture systems and lessons learnt in the recent past. In various chapters, the authors discuss innovations, approaches, and efforts that have helped overcoming gender-specific obstacles for mainstreaming women farmers in agriculture. I strongly believe that empowering women farmers by providing them a supportive environment and equal opportunities will lead to redefining their role as 'Agri-entrepreneur' than being remain obscured behind a masculine identity—'Farmer'. Equity-based solutions will empower them to strongly participate in decision making and establish them as champions of change to build a sustainable agriculture for feeding the world. I would like to congratulate both editors, Sugandha Munshi and Madhulika Singh, and all authors for developing this book that certainly will contribute to achieving SDG 2.

I wish to extend my sincere thanks and gratitude to the Springer staff, particularly Aakanksha Tyagi, Senior Editor (Books), Life Sciences, and Naren Aggarwal, Editorial Director, Medicine, Biomedical and Life Sciences Books Asia, for their constant support for the accomplishment of this compendium. The cooperation received from my senior colleagues such as David Morrisson, Peter Davies, and Daniel Murphy and my laboratory colleagues Abhishek Bohra, Anu Chitikineni, and Vanika Garg from Murdoch University (Australia) is also gratefully acknowledged. I would like to thank my family members Monika Varshney, Prakhar Varshney, and Preksha Varshney for their love and support to discharge my duties as Series Editor.

Perth, Australia Rajeev K. Varshney

Preface

> "Women farmers", a much ignored and many a times neglected terminology is the reality of the agriculture system in South Asia. Close your eyes, imagine a farmer what do we see, a man in the field, does a female holding a plough or using farming machines come to our mind? The answer is "No".

Women in agriculture often lack identity where either they are recognised as farmer's wife or a farm labourer. We often miss the obvious women farmers who contribute 60 percent into farm practices like sowing, transplanting, fertilizer application, weeding, harvesting, winnowing but are merely recognised and provided an equal level playing field.

There is an intrinsic need for the perception change as far as women roles and importance in agriculture are concerned. It is worrisome when despite their back-breaking work and contributions they are not counted. There remains an existential crisis for women in agriculture and gender data gap adds to the problem.

It's also perturbing as why agriculture has become a masculine synonym and how come this whole concept of agriculture got devoid developing understanding of agriculture being a man's business. Working with thousands of women farmers, every day they have taught us the patience to keep working hard without any expectation. Women farmers like Bholi Devi, Sumitra Devi, Kiran Devi, Anupa Devi, Laxmi Devi, Guddi Devi and many more have given us the chances to develop together a perception which brings reality to the forefront related to the immense capacity of women farmers, if they are provided with timely conducive environment and equal opportunity.

There are millions of women farmers working in agriculture system, in agriculture value chain and even then, they have a ghost identity. The challenges are many folds, patriarchal ecosystem to which she belongs, an ecosystem which refrains from allowing her to be recognised as a farmer.

But, while working with thousands of women farmers closely we found the hope, confidence and never-dying spirits of women at the grassroots who work from dusk to dawn relentlessly without any cribbing. They are always there to welcome you with open arms to learn, they are always ready to share their stories with a sisterhood approach and try together to redefine their identity. They are ready to grab the opportunity to progress. They are heterogenous group belonging to different caste, creed, region, religion but what is unique among each is the aspiration and power to dream. They think for their next generation, they work hard for their future, they want to create a better world for her girl child and hence for her farming is a way to survive and to also generate livelihood.

We strongly feel and believe *"THEY HAVE TO BE HEARD", "THEIR EFFORTS, THEIR HOPES, THEIR ASPIRATIONS, THEIR CHALLENGES AND THEIR JOURNEY NEED TO BE PENNED DOWN"*.

As a result, the book title speaks about the concept "Unheard being heard", not only the farmers themselves, but also the stakeholders working needs to be heard. Evidence needs to be shared. Sometimes small changes also have a deeper impact and in hush rush of achieving quantitative change we miss out on the quality side of any small effort and its impact. With the experience of women doing agri-business at the grassroot many a time there has been a hard-hit reality check, even if she makes a profit of $100 with some value-added work in agriculture, she continues with her girl child tuition fee to complete her studies.

These are hidden happiness which needs to be celebrated. Having said that, it does not mean that a meagre income generation is what we should aim for as far as economic empowerment of women in agriculture is concerned, but what we mean is that these small changes also act as a catalyst in strengthening her position in social structure and her self-esteem to become an agri-entrepreneur in coming days.

Along with the challenges, there is also a positive side to the story. The hope, aspirations, and efforts of many women farmers at the grassroot level who are silently working in their field, those whose voices might not be loud, but they are strong enough to bring a ripple in their patriarchal ecosystem, where with adopting improved technology on field they have started their journey as the champion of change. We also often forget to celebrate or better to say miss to hear the voices of these unsung women farmers, that also brings hopes for an equitable agriculture system.

An important intrinsic motivation to share the process, challenges, experience, strategy, and impact of the progressive women farmers is to highlight and understand what it takes to bring changes for achieving the goals of an equitable farming.

Though it's a long way for women in agriculture to be recognised but the journey has started, listen to the voices which are unheard and start being vigilant about the effort's women farmers are putting in ! We must recognise them and celebrate them. The self-help group is playing a crucial role as an innovation platform in increasing recognition and empowerment of women in agriculture system. In India for example, the livelihood mission launched by the Government is trying to nurture platforms for empowering women in agriculture. Bihar Rural Livelihoods Promotion Society--JEEViKA operational in the state of Bihar and Mission Shakti in the state of Odisha, are an example for mainstreaming women in agriculture

Hence, it becomes crucial to share stories and present facts about the challenges women farmers face and how they are making a way out of it. How the women farmers at the grassroots are taking one step ahead, what is the ecosystem support they require and what are the factors which keep them motivated for establishing their identity and creating a space in agriculture for themselves as farmers and not only as farmer's wives or farm labourers?

This book will put a dedicated effort to bring to the fore some of the unsung stories with facts, the efforts of women farmers in adopting sustainable agriculture practices and improved technology on their field, improved decision making and the triggering factors which add fuel into the life of a woman farmer to be an equal participant in the world of agriculture.

The chapters are written by the professional scientists working in agriculture with their own factual experiences and stories with the reality check of life of woman in agriculture and the change that they saw in the last decade.

The motivation behind writing this book remained the unsung stories of thousands of women farmers whom we have come across and how we start getting the scholars, practitioners, scientists, academicians working across South Asia sharing their experiences and together "drawing a rainbow" for women farmer they have worked with, the system they have strengthened, the challenges they have seen with women in agriculture as the first-hand experience and the efforts they have made together with women involved in agriculture to redefine their roles.

As far as any innovation or intervention designed for women farmers is concerned, it is important to understand that any system or personnel is able to contribute only because the women give opportunity to them to be a part of her journey. We are extremely indebted to each and every woman farmer with whom we have worked or with whom the authors of the chapters have worked.

While going through the book you will find that in each chapter along with the challenges discussed parallelly there are strategies, approach, efforts, and impact discussed. This book is not just about what is not working, this book is also about what is working and how it works and what is needed beyond what is being done. It is extremely important to share these unrecognized stories of women farmers who are bringing change at the grassroots. Let's celebrate the positive side, the small changes happening in the field of gender in agriculture.

The reality check and positive developments experienced by the experts, researchers, professional while working for and with women farmers need to be put together to highlight it is challenging to bring equity in agriculture, but definitely it is not impossible.

> Let's Celebrate Women farmers and women working in agriculture every day rather waiting for a Big Day to come. Let's Celebrate Her Identity!!

Patna, India

Sugandha Munshi
Madhulika Singh

Acknowledgement

Farmers who carry out arduous tasks and are backbone of our survival are our source of inspiration behind the conceptualization of this book. Experiences of working with various stakeholders, implementors, researchers, policy makers, scientists and farmers gave us the strength to come up with this book and explore the hidden realities. The contribution from the erudite authors of all the chapters made it possible to get our vision into reality.

We owe a debt of gratitude towards our working institutes and projects like Cereal System Initiatives for South Asia (CSISA) being a part of which provides a platform to systematically observe the existing dynamics on the topic. We are thankful to the editorial team of Springer Nature to accept the idea and for extending invaluable support throughout the whole process of publication. We are indebted to our parents, family members, colleagues and friends for providing a conducive space and time to contemplate this critical theme, which needs to be addressed in the present day. We are deeply indebted to thousands of women farmers who gave us an opportunity to be a part of their journey and continue to teach us.

We dedicate this book to millions of small and marginal women farmers across globe who are unheard but work hard from dawn to dusk!

Contents

About the Editors

Sugandha Munshi Lead Specialist - Senior Associate Scientist I has an intensive experience of more than a decade 12 years) working in the field of gender mainstreaming including research. She has extensive experience in providing strategic and technical leadership to government programs and projects, ensuring partnership between various key stakeholders in education and agriculture towards gender mainstreaming. She has received various national and international recognitions and awards related to her work in the field of gender and women empowerment like National Women Achievers Award India 2014, Young achievers Award 2018, Helrevo Women Achievers Award, IWD 2020, Inspiring Change Award, Planet 50-50 2020, Iconic Woman Leader creating better world for all by Women Economic Forum 2022, Gargi Excellence Award IWD 2023. She held the position of Global Shaper of the World Economic Forum, Geneva for consecutive 6 years and elected Hub Curator for two years. She has been an agenda contributor for women farmers at World Economic Forum. and she is also South Asia Ambassador for World Youth Social Enterprise, United Kingdom. Working in IRRI as a lead specialist, she has had various international as well as national publications on the work related to women farmers. With a system thinking approach she is working to mobilize science and knowledge by designing innovations to mainstream women in Agriculture and developmental outcome for farmers. She has had incepted the "Kisan Sakhi Model" for women in agriculture in Bihar, India focusing on Self Help Group as the conduit. She completed her Bachelor's degree from Miranda House and her Master's degree from Hindu College , University of Delhi. She holds her Ph.D. in Political Science with her dissertation on "Empowerment of Women Farmers : A study of Self-Help Group in Muzaffarpur and Motihari, Bihar, India".

Madhulika Singh Scientist and as an agricultural professional she has been working with CIMMYT for the last 9 years with an essence of agriculture research, extension and applying for all. Her experience as a research fellow has been with Indian Institute of Maize Research, Pusa Delhi where she did her PhD research work in affiliation with Banasthali University. She completed her Masters from Banasthali and Graduation from Gargi College, University of Delhi. Her inclination towards working at the ground level and for gender equality motivated her to be associated with the project for a long time. Her 10 years in the field of research has been versatile. She specialises in Maize Physiology and has done work on agriculture extension, tissue culture and molecular markers. The extension activities that she has been associated in CIMMYT includes sustainable agriculture, resource conservation, food security, farmers and extension workers training and have been coordinating with the National Agriculture Research and Extension Systems (NAREs). Her focus has been on research, extension activities related to balancing gender and this involves a lot of interaction with female and male farmers, understanding their requirements and then to provide technical inputs around those.

Chapter 1
Unheard Being Heard: Women Farmers: An Introduction

Sugandha Munshi and Madhulika Singh

> It's important for women to be heard, and to be heard its inherent to raise voices which needs to be loud and not sound like a noise. And your voice is loud when it is strong based on evidence

Pic Credit Sugandha Munshi for this chapter

S. Munshi (✉)
IRRI, Los Baños, Philippines
e-mail: s.munshi@irri.org

M. Singh
CIMMYT, Texcoco, Mexico

S. Munshi, M. Singh (eds.), *Women Farmers: Unheard Being Heard*, Sustainable Agriculture and Food Security, https://doi.org/10.1007/978-981-19-6978-2_1

Since the origin of the human civilization there has been a vocabulary bias: hu-MAN and it seems the civilization while evolving took it much more seriously than it should have and as a result everywhere, in every sphere of life—Man became a dominant decisive factorial figure.

Caroline Criado Perez in "Invisible Women," put it aptly when she says "Most of the recorded human history is one big data gap. Starting with the theory of Man the Hunter, the chroniclers of the past have left little space for women's role in the evolution of humanity, whether cultural or biological. Instead, the lives of men have taken to represent those of humans overall. When it comes to the lives of the other half of the humanity, there is often nothing but silence."

When we look at the agriculture and women, the story is not different. Infact, it is a burning example of how while the oldest profession of humanity evolved, the women roles were ignored, her efforts were silenced, and her identity was parked in a corner.

The ignorance where a mixture of the long-standing believes of viewing certain activities as the men business versus women business was one of the striking factors behind it. Sometimes the ignorance was deliberate and sometimes it was directed towards the power game. Simone de Beauvoir in her The Second sex very aptly put it—'One is not born, but rather becomes, a woman.'

Same applies in agriculture, "A farmer is not a man or a woman but just a farmer' whereas the social stigma and biased thought compartmentalized them and defined a man or a woman farmer."

1.1 Roads less Travelled

Across the globe, the challenges she faces are subjective and complex due to the cultural norms, lack of access, opportunity, and equal level playing field with that of men. In order to be more productive, they need access to resources and opportunities which is not readily available for them (FAO 2011).

When we delve deeper into the larger picture, we find women's contribution to be significant, their

efforts to be significant, their aspirations to be significant but what we do not find is a conducive ecosystem helping her to smoothly carry out her functions.

The agricultural labor force in developing countries comprises of 43% women varying from 20% in Latin America to closer to 50% in East, Southeast Asia, and Sub-Saharan Africa (FAO 2011). Around the world in all developing countries, they work as a farmer, as workers, and also as an entrepreneur. But across regions, they face inequalities and gender-specific constraints which acts as hurdles in their economic growth and productivity (FAO 2011). Women employed in agriculture also vary from region to region. In the Southern Asia region, since men migrate from rural to urban areas searching for better jobs, women are found 20.8 percent more likely to be employed in the agriculture sectors. But in rural areas, as a result of the absence of social infrastructure, the women are found to be struggling between paid employment and unpaid work (Beghini et al. 2019).

Specific to resources their access, ownership is meager. For example, when it comes to their control over land globally, they face the challenges where either they have no land rights or no decisions making on land, and often their tenure over the land is insecure. As far as their control over resources are concerned, the situation is grim. In developing countries, land rights are a decisive factor for their empowerment socially, economically, and even politically. When it comes to agriculture, it becomes more decisive factor which contributes to their overall development.

Women's land rights are a key determinant of women's empowerment in rural areas and have profound implications on women's ability to enjoy in practice civil and political rights, social and economic rights, as well as to escape poverty and social exclusion. Evidence shows a huge gap and gender inequalities in context of land in developing countries across globe. When it comes to agriculture, women having land ownership is less, women operating land is less. Their access to leased and rented land is less, and even the land they have are small and of poorer quality (FAO 2011).

The challenge she faces is multidimensional and hence the approach for the solving the issues also needs to be multipronged. When it comes to their education and use of improved practices in agriculture, modern technology, high-yielding seeds, improved fertilizers, pest controls measures, and mechanical tools, they are also found to be left behind (FAO 2011). The fact remains they are an active participant in agriculture, and we need conscious effort to mainstream her with improved knowledge and equal opportunities.

1.2 Women Are the Backbones of Agriculture!

Women play pivotal role to the success of agriculture as farmers, entrepreneurs, and laborers. Activities carried out by women range from small-scale collection and transport of water, fodder, and manure to other major agricultural tasks such as preparing the soil, sowing, and threshing. Women lack recognition as farmers, so they are refused of institutional bank support, insurance, cooperatives, and other government departments. Such situation needs to be addressed at both the policy and implementation levels. In Nepal the agriculture labor force largely consists of Rural women playing a vital role but yet struggle for recognition (UN Women 2017).

In India, the largest democracy in the world having the second largest population, its agriculture sector employs 80% economically active women comprising 33% agriculture labor force and 48% self-employed farmers (Oxfam 2018). In 85% of rural women only 13% own land. Rural women produce about 60–80% food. Women in agriculture are affected by identity crisis and no land rights. According to the general recommendation of the United Nations (UN) Committee on the Elimination of Discrimination against Women (CEDAW 2014) on the rights of rural women, "land rights discrimination is a violation of human rights." In the past, several steps have been introduced in this regard, proper implementation of which has remained questionable. Among the major challenges that curtail economic growth of women are tenure insecurity, no land ownership and inaccessibility to entitlements and no control.

Additionally, issues like improper care of land records, poor data management, and limited digitization of related records make identification of beneficiaries' difficult, thereby affecting implementation of agricultural schemes to uplift farmers. Forty percent of 505 women widowed by farmer suicides in 11 districts across Marathwada and Vidarbha, between 2012 and 2018, had no rights of the farmland they cultivated, 35 percent had secured the rights to their family house, and 33 percent of women did not know if they were entitled to a pension. Such studies reveal how women have been excluded from accessing institutional rights and entitlement as they lack recognition as farmers, Mahila Kisan Adhikar Manch (MAKAAM 2018).

According to 2015 census, in India, only 14% of women have land rights, whereas 86% farmers own less than 2 hectares of land. Among farming community, land is used as collateral security to get credit from banks which women hardly get to access. This scenario leads to less investment in farm and women-based enterprises. At present training is through cooperatives, Self Help Groups (SHG), and Farmer Producer Organizations (FPO) are being encouraged to train especially women farmers to achieve realistic growth of women and agriculture.

Human Development Survey (IHDS) states that 83% of agricultural land in the country is inherited by males. Enabling women farmers to have access to the same will have positive effects on the economy in and around the country. Provision of land ownership rights to women can be one of the important factors, contributing in raising total agricultural output in developing countries by 2.5–4% and has the potential to reduce global hunger by 12–17% Food and Agricultural Organization (FAO).

1.3 Women Roles in Agriculture: A Key Driver for Feeding a World Population of 9.3 Billion by 2050

By the year 2050, world will need to produce 60% more food for feeding population of 9.3 billion (FAO). Currently, worldwide almost 400 million women are engaged in farm-related activities. In total there is 43% women labor force in agriculture (Role of Women in Agriculture) ESA paper 11–02, March 2011, FAO (www.FAO.org/economic/ESA). Their role is inevitable in non-mechanized farm occupations including sowing, winnowing, harvesting, and other forms of labor-intensive processes such as rice transplantation and increasing their role in mechanized farming is the need of the hour. Ignoring them now can be dangerous as far as hunger is concerned. Women's labor contributions in agriculture are almost half, ignoring them, their need and not supporting them in the progression will act as an important hurdle in feeding the world

Myths Which Need to be Shun Off before it's Too Late

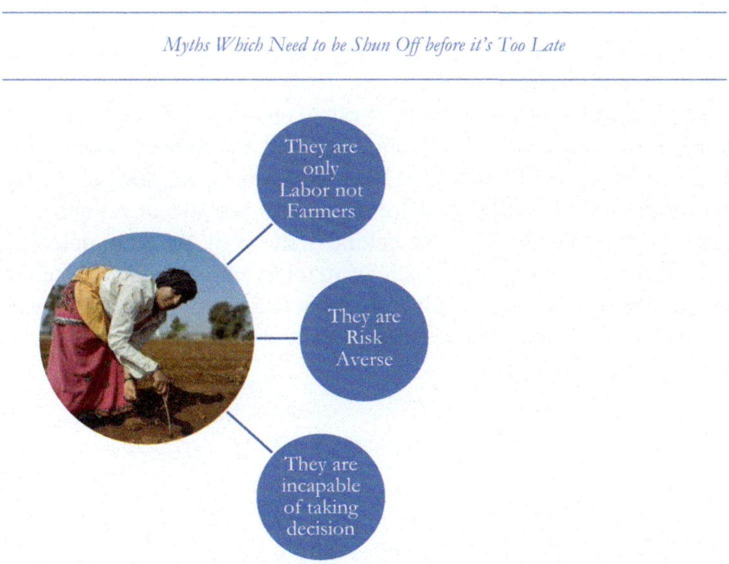

The myths are the result of the biased thinking developed over a period of life span. This is engraved in our subconscious mind where due to the patriarchal ecosystem generally the thinking capacity of an individual get prisoned. Power dynamics prioritizing men over women is engraved which always leads to an unequal distribution of resources and opportunities for women.

As a result of which we often ignore the possibilities around. Many a time as an implementor, development practitioners, policy makers, scientists, and as farmers, we often ignore the established reality "A woman farmer is the backbone of agriculture" and ignoring them will create a hollow system aiming for improved farming and feeding the world. As a result of the biased lens, we forget to involve them in agriculture reform, we ignore half of the population and very easily avoid listening to them and their needs.

We often get confused with her multidimensional roles as her weakness and reasons to ignore her. We easily make statements "ah they cannot participate in the training because they are women; reality being we cannot design a training which incorporates her need and is women friendly; ah they cannot adopt technology because they are women;reality being we do not take or design technology keeping her at centre; ah they cannot take risk, reality being we do not create a conducive environment where she can take risk with technology; ah they cannot drive operational machineries for farming; reality being we fail to design machines keeping her need at focus or as far as machineries is concerned better to say we forget her completely; ah they are not farmers,they are farmers wives reality being they are farmers and we are gender blind, ah they cannot be a service provider for agriculture practices reality being we fail to provide them adequate knowledge and education on financial literacy ....The statements goes on and on!

The challenges are many folds, and with the challenges, there are equal opportunities to set examples nullifying the biases and establishing evidence-based development stories and efforts trying to fill the gap and being closure to the reality. Believe women across globe at the grassroot levels are doing back-breaking work on farm, they are adopting improved technology, they are doing varietal replacements, they are establishing their identity as farmers, they are involved in agri-entrepreneurship, they are becoming service providers, though the number are few, but it is only with a convergent approach and a dedicated effort with unbiased approach these numbers can increase. To delve deeper, we need to explore factors which contribute into the same. If there are certain positive developments what are the reasons and supportive factors behind that, is it the self determination of individual women farmers, is it the increasing interventions being piloted by various stakeholders like government, policy makers, farmers, scientists developmental practitioners, etc.. ? Let us explore them and learn from them !

1.4 Missing on Women Data Is Missing on Reality

While resolving any issue data is the key. Data gives us the rational and basis to design intervention aiming towards bringing solution to the gap and challenges. As far as women in Agriculture goes this is one of the perturbing issues, "Lack of Data of women in agriculture."

While any survey is planned by researchers, it is important to include the women respondents of the household. Mostly while designing of the study and surveys, we miss out on the other half of the population, which is involved in agriculture and at the end of the analysis what we get is a gender-blind data and start running from pillar to post. Is it even required when we are sensitive enough during the designing phase itself? Ask Yourself. Data which is devoid of equity is devoid of reality, it cannot be only related to men. In order to get a complete picture we must fill the gender data gap. In one of the report published by Bill Gates in 2012, they mentioned how while doing the assessment of a breeding variety an organization only spoke to the men farmers and at the end while adoption was concerned households did not adopt it. When it was explored later while missing on women first, it was found there were several factors considered by women which did not come earlier, like the cost of land preparation, weeding the duration of cooking of the crop and hence unsurprisingly there was no adoption (Criado Perez 2019). Women farmers are ready and willing to share their thoughts, experiences, and opinions, the question with us is "Are we willing and ready to listen to them" and generate evidences"?

1.5 Rising Hopes: One Step Ahead as Farmers!

Well having said that and with innumerable evidence of finding women left behind in agriculture system where they are continuously struggling for establishing their identity as farmer, availing equal opportunities, equal rights and above all an equal level playing field with men to contribute to the agriculture, there are high hopes which reflects that the journey for equity in agriculture is a long way even in today's twenty-first century but relentless of that, "a conscious dedicated effort" resolving these hurdles is the only solution. Women are also found participating in the various forms of processing and marketing of agriculture produce, along with the cultivation (Aggarwal 2003). Self Help Group in India has played a decisive role in redefining role of women in agriculture. The group affiliation provides opportunity for women in agriculture for trainings, exposures to improved farm technology and also creates a conducive environment where they can negotiate and take risks for agri-entrepreneurship (Munshi and Bhardwaj 2021).

Two Generations Women Farmers, Munger Bihar India

We cannot sit back being a mere spectator and wait for a time when things will improve on its own. What is required is dedicated efforts from all the stakeholders involved in bringing equity in agriculture. Right from farmers both men and women at the farm, public-private system involved in agriculture, agriculture universities, scientists, research institutes, government, civil society, opinion builders like media and the citizens needs to come together.

There are positive developments on the field which supports the need for the agencies to come together and work for women farmers. Though the challenges are immense but slowly and steadily there are dedicated efforts at government and various agencies which has started the story of shifting the biased reality.

Consortium of groups in Agriculture reform is such an example which has globally started talking about gender more seriously than ever before. The governments in South Asia have started dedicated plan of actions for mainstreaming-Women in Agriculture. At least there is an increasing buzz around in the last five years focusing on gender in agriculture and how to make the agriculture system inclusive. Donors have become cautious of their investment to promote equity.

It is indeed an interesting observation working closely with policy makers, farmers, donors, and various stakeholders in the field of agriculture where it is found that there is an increasing ownership of change as far as the role of women and their recognition is concerned. This is a positive signal but equally there has to be a dedicated plan of actions and implementation of those actionable points. Its a long journey but at least the caravan has started. And yes, while working for the development of women farmers, we need to share their voices, their stories as raw as possible. Each and every individual women farmers has their own respective journey, generalizing them can devoid the reality.

1.6 Walk the Talk

Along with the gap identifications, it is equally important to document the stories, interventions, innovations, and efforts made by farmers and aligned agencies which are working for improving the farmers livelihoods focusing on women farmers.

This particular book is focused with the objective to highlight the challenges and efforts, interventions and innovation being implemented across South Asia at the grassroots by farmers, policy makers, scientists, and implementors for mainstreaming women in agriculture. This particular book is an effort for "Making the voices Unheard being heard." The voices which are working on field, at policy, at implementation and at the innovation level are being covered in the book which is based on firsthand experience rather rhetoric. There is a blend of stories which is evidence based highlighting the challenges and efforts made in the current existing challenges on the pathway of empowering women in agriculture. There are nuance critical observations of the journey of women farmers establishing their identity and how various stakeholders are working with the mission of mainstreaming women in agriculture. In developing countries agriculture is the backbone of economy and most of the population directly or indirectly are affected. Women are playing crucial role in the agrarian society, they need to be celebrated and their stories, challenges, efforts, vision and expectations need to be catered by government, researchers, public sectors, and social welfare programs. Hence, while we try to bring reform as far as women in agriculture is concerned, it is non-negotiable not to involve the voices, opinions, and needs of the women. The progressive effort since a decade calls for an urgent documentation which highlights and talk about the efforts being made on field by women farmers, agriculture department, aligned government departments, research institutes, development agencies etc working with women in agriculture.

In this book, a conscious effort is made to bring the unsung stories to the forefront with data. World hit by Pandemic Covid 19 brought the hidden reality of how women in agriculture are more vulnerable when any crisis stuck and hence its again a reminder "we need to reach out to women farmers directly," through innovations, strategic planning, and digital world. This book is a sincere effort to bring to the fore some of the hidden realities which are adding fuel to the life of a woman farmer and the existing challenges which needs to be sorted out for her to be an equal participant in the agriculture domain. The chapters are written by professionals, scholars, implementors, scientists working in agriculture with their own factual experiences and stories with the reality check of life of woman in agriculture and the change that they saw in the last decade.

> Women voices are crucial to address climate change, for a sustainable agriculture practice and for developed economy, don't ignore "Hear the Unheard"!

References

Beghini V, Cattaneo U, Pozzan E (2019) A quantum leap for gender equality: for a better future of work for all. International Labour Organization

Criado Perez C (2019) Invisible women: data bias in a world designed for men. Harry N. Abrams

FAO (2011) The state of food and ambassador, women in agriculture closing the gender gap in agriculture. FAO

Munshi S, Bhardwaj P (2021) Self help groups - a catalyst for empowering women in agriculture! EEO 20:5404–5412. https://doi.org/10.17051/ilkonline.2021.01.574

Oxfam India (2018). Move over 'Sons of the soil': why you need to know the female farmers that are revolutionizing agriculture in India

UN Women (2017) Women farmers of Nepal take charge of their lives and livelihoods. https://www.unwomen.org/en/news/stories/2017/2/feature-women-farmers-of-nepal-take-charge-of-their-lives

Chapter 2
Mainstreaming Women Farmers: Innovations and Approach

Sugandha Munshi

Abstract Mainstreaming women in agriculture is the need of the hour and its imperative to bring and discuss innovations and approaches which can accelerate the pace of it. In spite of her back-breaking work women farmers remain invisible and is aspiring for her identity as a farmer. There are various challenges which is in the ecosystem of a woman farmer ranging from lack of an equal level playing field with that of men, identity crisis, lack of gendered extension system, lack of control over resources like land, input etc. which needs a planned strategy and intervention aiming to reduce the gaps. There are innovations at the grassroots operational which helps in creating a conducive environment for women in agriculture to initiate and participate in the agriculture development. Self Help groups (SHGs) working in India have brought a momentum to the process of mainstreaming women in agriculture with the provision of a platform led by women to learn new technologies, implement the learnings on their farm and also creating larger scope for women-led entrepreneurship in farming's. At the grassroots, village level an inclusive extension system where a right number of women officials works helps in reaching out to women farmers more easily owing to the social norms where interaction between women is an acceptable norm rather men interacting with women. In the world of new normal now, "Information, Education and Communication" (IEC) materials have a huge role to play as far as instilling the right approach towards tackling the gender biases existing in the field of agriculture is concerned. As far as the communication tools and materials are concerned, we need to pull our socks for making it inclusive.

Keywords Women · Agriculture · Equality · SHGs · Inclusion · Strategy

---

S. Munshi (✉)
IRRI, Los Baños, Philippines
e-mail: s.munshi@irri.org

S. Munshi, M. Singh (eds.), *Women Farmers: Unheard Being Heard*, Sustainable Agriculture and Food Security, https://doi.org/10.1007/978-981-19-6978-2_2

2.1 Introduction

Pic credit: Sugandha Munshi

Close your eyes! Imagine a farmer, what is the image you got? Is not that a man with a turban and with an axe on his shoulder and a woman standing beside her as support? That is supposedly quite a natural phenomenon, but is it so? When we imagine farmers, mostly we associate them with a man rather a woman. And this is not just a rhetoric rather the perception developed over a period of time where a farmer means "a man." It is important to first accept that there remains a huge gender gap as far as agriculture is concerned and women farmers are many a times left behind. Data gap is one of the critical obstacles as far as women's contribution in the agriculture is concerned (Criado-Perez 2019).

Although she has a significant contribution in the agriculture at every stage and she does farm enterprises in many developing countries, still there remains a huge gender gap due to various constraints operational in her ecosystem (FAO 2011). Women comprises of 43% of agriculture labor force in developing countries and work longer than men. Inspite of playing key roles in various agriculture activities her access to resources and opportunities are less than that of the men (FAO 2011). At global level, women access to extension system is less than 10%. To make agriculture sustainable, it would be important to close the gender gap existing in the agriculture system, which will be beneficial for both women and agriculture (FAO 2010-2011). Closing the gender gap would play a twin role of promoting gender equality and creating an opportunity for economic development where women plays crucial role in economic development through agriculture (Quisumbing et.al. 2014).

2.2 A Level Playing Field for Women in Agriculture! Challenges

Pic credit: Sugandha Munshi

Both men and women are equal participants in the agriculture activities. It is important that both get an equal opportunity and level playing field. It has been a long journey for women since the human evolution to mark her presence in the patriarchal social structure. A normal reality is often ignored on the pretext of culture, traditions or better to say illusioned reality when women are taken as a secondary citizen in the evolved society.

As far as agriculture is concerned, the story is not very different. It is extremely important to understand the female association with the place of her belongingness has a direct impact on the opportunity she has. Many a time the opportunities are scarce owing to the rural area they belong to. In agriculture their access to resources is not equal to that of men. Had the women enjoyed the same access to productive resources as men, women could boost yield by 20–30%; raising the overall agricultural output in developing countries by 2.5–4%. This gain in production could lessen the number of hungry people in the world by 12–17 percent, besides increasing women's income (FAO, 2011).

It is also important to realize that women often have no land rights to which they are more vulnerable towards availing equal opportunities for improved farming. There is a wide gap as far as her association with the land is concerned. Land is critical as far as the livelihood of millions of women in Asia, Africa and Latin America is concerned. Comprising around half of the total agriculture labor force, still less than 15% of landholders around the world are women (FAO 2018). At the grassroots, while women do not have land rights, it creates hurdles on the way of their own decision-making for practicing farming in the way they want. There are also significant social and economic impacts related to the denial of women right to land (Agarwal 2003). In developing countries, between 10 and 20% of all landholders are women. Worldwide, women access less than 10% of agricultural extension services. After the food price crisis of 2007–08, women-headed households were 1.6 times more likely to be food insecure than those headed by men (FAO 2018).

2.3 An Invisible Entity: A Woman Farmer!

Identity is another key area of concern as far as women and agriculture is concerned. Issues related to land rights of her is one of the significant hurdles behind the concrete identity establishment of women in agriculture. Often, they are recognized as Farmers' Wives rather than Farmers themselves. There is often a strong relation established ideologically of the word farmers as someone with masculine identity (Margreet: Thrive). Farmers are a masculine figure in the subconscious mind and majorly such impression is formed due to the patriarchal ecosystem around.

The issue which is pertinent is the absence of women farmer's recognition as the farmer. Across globe 80% of the rural smallholders' farmers are women and produce 50% of the world's food (Action Aid 2011). A harsh reality but true to its core—"-She is invisible "on the fields, on the papers and many a time in the policy reform." Often considered to be "Farmers Wives" she loses her identity of being a farmer. The gender-blind system has much to contribute into it. A gender-blind ecosystem creates an environment where inspite of doing back-breaking work her efforts remain overshadowed by the patriarchal biased thought of farming to be associated with men. At household and community level who will be considered as farmers is directly influenced by the gender norms. They remain generally unnoticed or are underestimated as farmers by men and women of household and community. Women are believed to be just a supporter of Farmer who are by default men (Galie et al. 2013).

It is important to delve deeper into the identity aspect which in turn leads to the gender gap in agriculture at larger level. Often women working on the field are a labor. Also, a perception of farmers who take lands on rent to do farming has extreme contradictions. Ironically while working with farmers on field, it is found, that men taking land on rent to do the farming are farmers but when it comes to women doing farming on a rented land freezes her into the category of labor.

It is important to understand in this, that the identity of women as farmer is close to them. Women in villages feel proud of being associated farmers and look forward to being called as a farmer. A group of women farmers working with Jyoti Mahila Federations in the state of Bihar at District Muzaffarpur incepted their own identity as the "Kisan Sakhi" in the year 2014. It is equally important that when it comes to the identity, it does not happen in isolation but equally with the tangible efforts made on the field. Bholi Devi one of the Kisan Sakhi "Along with the opportunities, new technology information on improving our productivity in wheat and rice our knowledge has acted as a catalyst in changing the biased perception of our identity as a farmer in our community. We are taken seriously and are recognized as informed and progressive farmer (CSISA 2016).

There are certainly more dialogues and evidence generation required to set up a correct connotation for her identity and rights as a farmer.

2.3.1 Inclusive Extension System: Key for Sustainable Change

Pic credit: Sugandha Munshi/CSISA: Mahila Samkhya Extension officer addressing the trainers

Culture and traditions are more closely associated with women in the social arena. It is also a challenging reality as far as women in agriculture is concerned. Women farmers are comparatively less mobile as compared to the male farmers owing to social norms, traditions, time constraints and the more responsibilities they must undertake at the household level. There are cultural and social obstacles which keep their movement outside their village and home restricted (FAO 2011). When we delve deeper into this, it is often found that even if the women farmers want to access knowledge, the non-conducive extension services act as an obstacle. The interaction with men extension agents is often found to be not efficient to the social ecosystem she belongs and increases the chance of her being left behind in gaining awareness and knowledge on improved farming. Many women farmers are more comfortable to talk to women extension agents than men. To mainstream women in agriculture, inclusive extension system is the most crucial element to put into consideration

"How do we make our extension system gender sensitive and gender inclusive"? The answer is a "Gendered Extension System", where extensions officials are gender sensitive as well inclusive in their approach and numbers.

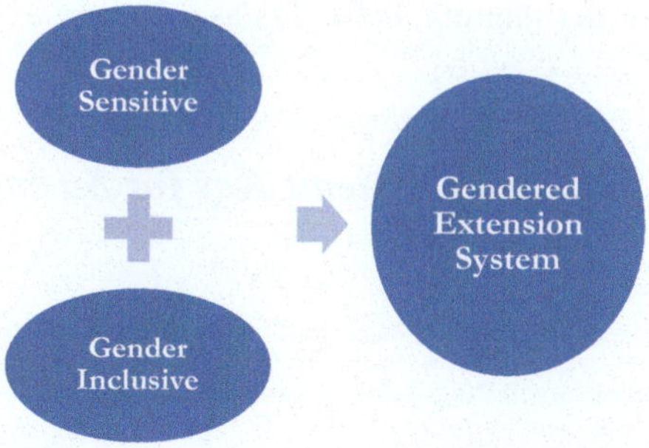

The challenges faced by women farmers in agriculture are also related to the opportunity gap they face owing to biased extension system putting in consideration the needs and constraints of women farmers. There is a planned strategy which needs to be incorporated in the extension system making it more women friendly and catering to her needs. The basic requirement of which incorporates including more women as the extension personnel, so that from the very beginning, the women farmers feel comfortable to interact with them. Presence of women personnel act as an ice breaker for them to be comfortable and approachable. Many women farmers feel comfortable to share their problems with female extension agents (Due et al. 1996).

Pic credit: Sugandha Munshi—Women extension officials planning

Incorporating a planned strategy to open space for more and more women extension personnel needs to be introduced into the institutional structures working for farmers. Along with this training men personnel to be gender sensitive and follow inclusive approach while working with farmers is an important requirement. A gender sensitive man who is an extension official helps in the ice breaking of the complete biased thought at the community level that reaching out to women farmers is only a woman business equally it establishes the understanding among the women farmers "learning must not get affected by who is there to bring knowledge to you, a man or a woman what matters is the value of the knowledge which will empower you with the weapon of knowledge and awareness." This is a long process which take place over a period. And to bring that sustainable equity in agriculture small steps like an inclusive extension system will act as the steppingstone. A right balance of both men and women extension personnel is needed. It is possible that many a time having equal number of women extension personnel is not possible but at least a certain percentage have to be made mandatory. Along with that gender sensitive training to men extension officials/personnel is crucial where they can deliver sensitive and qualitative services (FAO 2011).

2.4 Farm Mechanization: Need for Gendered Greasing!

Working for women farmers and working with women farmers might seem to be the same but there is a subtle difference between both. When we work for women farmers, there is a high chance that the solutions we propose are standalone and far away from reality.

Pic credit: Sugandha Munshi—Jyoti Mahila Samakhya Federation's Women farmers at Bandra, Muzzafarpur Bihar with their Machine Rice Transplanter

But when we work for and with women farmers the solutions proposed, and problems understood is closer to reality. Women farmers are very much open to mechanization, where they prefer it owing to less drudgery and time-saving factors attributed to its operational side. But designing any blanket technology recommendation for improving the productivity, saving labor, and saving cost is "Not" the solution. It does not cater to the needs of women farmers. The machines developed must include women needs as the pre-requisite. The developers of farm machineries must be careful and aware of "Labor saving device" or "Male labor-saving Device"? A gender-aware and a gender-responsive lens are necessary for farm mechanization.

Mechanizing of the agriculture task in Turkey resulted into the decrease in women participation in agriculture activities and when it was scrutinized, it was found that the machine designed could not be used by women, because it was not designed including her needs (Criado-Perez 2019).

Designing technology for women farmers requires intensive time and intellect investment with them. Now when we have plethora of such machines around us what to do? The answer lies again in re-thinking over the existing farm machineries to be made useful for women.

For instance, the whole concept of Service provision can be interlinked with the heavy machineries like Zero Tillage of wheat, machine transplanter of rice, etc. as a business model for her? Why not? They are ready to do it. The question which I wonder, when service provision can be done by men then why women cannot do the same? They Can.

What is required is her financial literacy and a deeper understanding of business model same as that of men, rather than getting only into the needed debates of the machinery not being conducive to women farmers.

Pic credit: Sugandha Munshi—Planning by women farmers, yoti Mahila Samakhya

Having said that there is an urgent need for developing such cost-saving machinery which is beneficial for women farmers and can be easily used by them. I leave this to the scientist community of Farm Engineers! They must be gender sensitive and think and act on how once we design the farm machinery "Can that be introduced to the whole population of farmers which includes women too"? If the answer is no, remember it will not be sustainable.

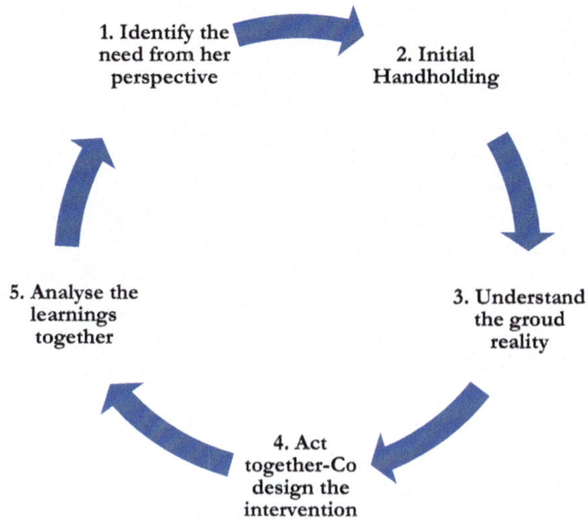

Model of working with women farmers

2.4.1 Self Help Group as the Conduit for Improved Extension System: A Silent Revolution for Women Farmers Identity, Awareness, and Empowerment (IAE)

To deal with how to approach women so that they get equal access to knowledge and opportunities designed for improving their productivity and, income, reaching out to the women farmers through the Self-help Group is most impactful. The self-help group is an association of women coming together. In India SHG has become a movement of women for their empowerment. It is an important platform for women to come together and create their space in social, economic, and political arenas.

It involves a bottom-to-up participatory approach for empowering women economically, socially and politically. It is like a small group of voluntary association which largely consist of women 10–20 in numbers coming together to solve their common issues through drawing a common strategy. The exclusive aspect of the intervention to solve any issue be it social or political is owned by the members there is an intrinsic motivation among the members to bring a solution and hence the change brought become sustainable.

In India, self-help groups (SHGs) have played an important role in changing the lives of women in rural areas and are therefore considered one of the most significant tools in implementing the participatory approach towards the economic empowerment of women and in improving various aspects of the social structure in the country. These groups, which are formed and usually supported by nongovernment organizations (NGOs) or, now increasingly, by government agencies, are small voluntary associations of poor and marginalized people, preferably from the same socioeconomic background, whose structures, processes, and activities provide their members with the opportunity to identify for themselves the problems that confront them and seek the solutions that they can and are willing to implement. These also provide their members with better access to support services, including credit and government extension services. By empowering rural women this way, SHGs have become the vehicle of change for poor and marginalized people to be released from the clutches of poverty (Sahu and Singh 2012), while enhancing the status of women as participants, decision-makers, and beneficiaries in the democratic, economic, social, and cultural spheres of life (Kondal 2014). Women's participation in SHGs has created tremendous impact upon the lives of poor women and has empowered them at various levels, not only as individuals but also as members of families, and communities. It creates a conducive platform for the members to take risk Being in group, for the extension people to reach out to women farmers become easy have a ripple effect vis a versa it also give confidence to individual woman to come out of their household and participate in agriculture extension programs and trainings (Munshi 2017). This has become a successful tool for women at the grassroots of India to get rid of poverty along with their empowerment.

Self-help group also provides a friendly/ favorable platform for participatory approach for empowering women socially, economically, politically as well provides them with a level playing field to take risk. Being in group an individual women get chance to share risk and adopt innovation in their work arena. The following picture needs to be replaced with the one attached.

Especially in agriculture, where through the platform they get trainings, credits, and opportunity to learn new technology. It is also important to realize that being part of a group helps an individual to come out of their cocoon and fellow women in the group act as a strong support system. In one of the project Cereal System initiatives for South Asia, there has been a first-hand experience, in the state of Bihar where working Shardha Samooh, SHG of Jyoti Mahila Samakhya federation, the group learnt the Mechanized transplanting of Rice technology and bought their own machine becoming the first women group of service provider in the region. Service provision in agriculture is believed to be a male dominated enterprise but the women group taking initiative of becoming service provider not only challenged that notion but also established an example at the grassroots. Additional to increasing the productivity, reducing the cost of input this woman led service provision enterprise provided them with higher income opportunity and also reducing drudgery.

The SHG platform is helpful for bringing women at the village level together where they intensively discuss the issues affecting their social, economic and livelihood arena. Through this platform their risk taking capacity increases as a result of the group affiliation and increasing power to negotiate (Munshi and Bhardwaj 2021). Not only this it also act as a strong mechanism for reaching out to women as direct stakeholder for developmental intervention and as far as agriculture is concerned, it is a strong operational mechanism for innovation in farming. When the women involved in agriculture come together, first it creates a positive environment of sisterhood, a safe space where they can brainstorm together their own subjective needs and aspiration and take initiative thereby to try, practice and then finally adopt any technology. In last 8 years of on field and policy level initiative. I found the intrinsic motivation to own change by the women farmers is the key. Any external motivation can start a change for a while but to make it sustainable what is required is the ownership of the women farmers of her own development. This comes from the experience of working during inception of Rice Nursery Enterprise as a viable option for increasing women farmers' income. One of the hundred of farmers who started the nursery enterprise in a group, Sumitra Devi of Harpur village at Muzaffarpur in the state of Bihar started her own venture along with her husband back in 2015 and still till date she continues to do so. The learning from this put light on the often ignored reality, once you create awareness about certain technologies, women vet them according to their own subjective farm needs whereby they adopt the technology with the permutation combination of the benefits it brings and the impact it has, in her own area of influence which involve households, family, community, etc. Equally with progressive time, now it is crucial to strengthen the Self Hep Groups as a platform providing information's which helps in changing the dynamics of women's participation in agriculture to impactfully translate information gained and knowledge acquired into the implementation (Raghunathan et al. 2018).

2.5 Making Invisible Visible: Unavoidable Need for Second Green Revolution!

Pic credit: Sugandha Munshi

The invisible "women farmers" act as the backbone of the innovation in agriculture system. Not only in India but across the world, women's contributions to agriculture are significant. The UN Food and Agriculture Organization estimates that if women had the same access to productive resources as men, they could increase yields on their farms by 20–30%. This could raise total the agricultural output in developing countries by up to 4%, which could, in turn, reduce the number of hungry people in the world by 12–17%—that is 100–150 million people. About 60% of agricultural operations like sowing of seeds, transportation of sapling, winnowing, storage of grain etc. are handled exclusively by women, while in other jobs, men share the work with women. Apart from participation in actual cultivation, women participate in various forms of processing and marketing of agricultural produce (Agarwal 2003).

Hence, it becomes crucial to initiate inclusive approach right from policy to implementation level where women farmers are mainstreamed and made visible (Munshi 2017). In India, the announcement of October 15th as the women farmers day by government is an important landmark towards mainstreaming women in agriculture and making her efforts visible. This has a huge scope of sending positive messages at the grassroots of the country were acknowledging her contributions can start a momentum on the field as farm as equitable farming is concerned.

However, as far as the communication tools and materials are concerned, we need to pull our socks for making it inclusive. In the world of new normal now, "Information, Education and Communication" (IEC) materials have a huge role to play as far as instilling the right approach towards tackling the gender biases existing in the field of agriculture is concerned. It became far more relevant where the digital world is having a huge say in building public opinion and influencing their thought perspective. Most of the IEC tools seems to overlook equal representation of women farmers. There are often male farmers to be seen when we go through media and communication related to government announcements on farmers welfare program, subsidies announcements, policy updates, etc.

Pic credit: Sugandha Munshi: Women Farmers in a meeting with Jyoti Mahila Samakhya Federation, Bihar India

It is extremely crucial to develop a directive policy from Government of India in agriculture system to follow proper protocols while designing and disseminating IEC materials where there is equal representation of women in communication materials like pictorial publications, animations, caricatures etc. Along with that while referring farmers in any address or announcements we address women farmers as "Kisan Behno," the same way refers to men farmers as "Kisan Bhaiyo." Finally, inclusion at all levels will accelerate the prospect of achieving second green revolution in India since women are the backbones as far as agriculture and rural India is concerned.

References

Agarwal B (2003) Gender and land rights revisited: exploring new prospects via the state, family and market. J Agrar Chang 3(1&2):184–224

Action Aid (2011) What women farmers need: a blueprint for action

Criado Perez C (2019) Invisible women: data bias in a world designed for men. Abrams

Due JM, Flavianus M, Temu AA (1996) Smallholder farmers' perception of extension by gender in Tanzania. Sokoine University of Agriculture, and the University of Illinois, Morogoro

FAO (2011) The state of food and agriculture 2010–2011. Women in agriculture: closing the gender gap for development. Food and Agriculture Organization of the United Nations, Rome

FAO (2018) Gender Gap in land rights 2018. https://www.fao.org/3/I8796EN/i8796en.pdf

Galiè A, Jiggins J, Struik PC (2013) Women's identity as farmers: a case study from ten households in Syria. NJAS Wagening J Life Sci 64–65:25–33

Kondal K (2014) Women empowerment through self-help groups in Andhra Pradesh, India. Intl Res J Social Sci 3(1):13–16

Munshi S (2016) CSISA-Trained Women Farmers Lead the Way for Agriculture in Bihar. https://csisa.org/csisa-trained-women-farmers-lead-the-way-for-agriculture-in-bihar/

Munshi S (2017) Time to recognize women farmers in India. World Economic Forum. https://www.weforum.org/agenda/authors/sugandha-munshi/

Munshi S, Bhardwaj P (2021) Self-help groups - a catalyst for empowering women in agriculture! EEO 20(1):5404–5412

Quisumbing AR, Meinzen-Dick R, Raney TL, Croppenstedt A, Behrman JA, Peterman A (2014) Gender in agriculture: closing the knowledge gap. Springer

Raghunathan K, Kannan S, Quisumbing AR (2018) Women's self-help groups, decision-making, and improved agricultural practices in India: from extension to practice. IFPRI discussion paper 1735. International Food Policy Research, Washington, DC

Sahu L, Singh SK (2012) A qualitative study on role of self-help group in women empowerment in rural Pondicherry, India. Natl J Community Med 3(3):473–479

Chapter 3
Women Farmers or the Farmers' Wives: Unveiling the Negotiated Gender Roles

Sujata Ganguly

Abstract The male dominance in decision-making in the household and economy has continued even in areas where women are the key providers of labour because the influence of women has not been recognized. The broad objective of the study is to understand the similarities and differences (if any) between the lives of women farmers and farmers' wives in terms of decision-making and economic independence. The 2015–16 National Family Health Survey (NFHS-4) data has been used. The available variables from the datasets were used to define the women farmers and the farmers' wives. This study raises several questions which are just the tip of the iceberg. Delving deep in this subject unveils enormous unknown facts that are beyond the purview of our own horizons. 'Gender' is a subject where more than numbers, details matter. There is no denial of the importance of numbers because that depict volume but it resembles an onion where we peel and unveil layers after layers before we get to the core. Nevertheless, not all women are marginalized or disempowered. The decision-making, mobility, and conjugal relations are determined by the socio-demographic factors and the existing social and cultural norms. At times, we accept and submit to a culture believing it as the truth of life. The study demands a thorough qualitative study to answer some of the questions and question some of the facts.

Keywords Household · Decision making · Woman · Farmers · Decision-making · Dominance

3.1 Introduction

As has been underscored by the extensive and growing body of literature on gender and agriculture, women's roles and responsibilities in agriculture-within the household and the community-are complex, diverse, and multifaceted (Doss 2001; Doss and Morris 2001; Meinzen-Dick et al. 2011; Quisumbing and Pandolfelli 2010).

S. Ganguly (✉)
Principal Researcher (GESI), Includovate, New Delhi, India

S. Munshi, M. Singh (eds.), *Women Farmers: Unheard Being Heard*, Sustainable Agriculture and Food Security, https://doi.org/10.1007/978-981-19-6978-2_3

Women play a significant and crucial role in agriculture, its development and allied fields like the main crop production, livestock production, horticulture, post-harvest operations, agro/social forestry, fisheries, etc. The nature and extent of women's involvement in agriculture, no doubt, varies greatly from region to region. Even within a region, their involvement varies widely among different ecological sub-zones, farming systems, castes, classes, and stages in the family cycle. Studies suggest that, despite the wide variability among regions and countries, women farmers play a major role in labour-intensive activities such as planting and weeding, among many other agricultural activities (Meinzen-Dick et al. 2011). With the exception perhaps of ploughing, women are actively involved in almost all agricultural production work. It also reveals that female time use in agriculture varies widely depending on the crop and the phase of the production cycle, the age and ethnic group of the women in question, the type of activity and a number of other factors. Time use studies (Bardasi and Wodon 2009) have shown that women not only have substantially less free time than men, but are more often confined to performing time-intensive and socially unrewarding activities such as fetching water and fire wood.

As far as the decision-making is concerned, male dominance in decision-making in the household and economy has continued even in areas where women are the key providers of labour because the influence of women has not been recognized. To reword Doss (2001) but in another regard, the majority of women farmers live in male-headed households and farm on male-owned plots. Access to resources does not guarantee control over them (Ngenzebuke 2014) since the one in control might dominate in the decision-making. There seems to be an emerging opportunity for women to access the market. However, the scope for further research remains whether their access to market is limited to physically going and selling the products without having any control on the income. As pointed out by Kabeer (1999b), we can also distinguish between various critical 'control points' within the decision-making process itself where such control is defined in terms of the consequential significance of influencing outcomes at these different points. As cited in Kabeer (1999b) that Pahl in 1989, distinguishes between the 'control' or policy-making function in making decisions about resource allocation and the 'management' function, decisions which pertain to implementation.

Introspecting deeper, households are also fluid or flexible. 'Statistical' perspectives on decision-making, however, should be remembered for what they are: simple windows on complex realities. They may provide a brief glimpse of processes of decision-making, but they tell us very little about the subtle negotiations that go on between women and men in their private lives. Consequently, they may underestimate the informal decision-making agency which women often exercise. The formal account of decision-making given by women ascribed most of the power to men: the husbands were said to be 'heads' of households and their 'owners'. There is an important body of research from the South Asian context which suggests that the renegotiation of power relations, particularly within the family, is often precisely about changes in informal decision-making, with women opting for private forms of empowerment, which retain intact the public image, and honour, of the traditional decision maker but which nevertheless increases women's 'backstage' influence in decision-making processes (cited in Kabeer 1999a). Based on the discussion above, the broad objective of the study is to understand the similarities and differences (if any) between the lives of women farmers and farmers' wives in terms of decision-making and economic independence.

The literatures also suggest that the women involved in rice cultivation in India were observed to participate in transplanting; pulling of the seedlings; weeding rice paddies; harvesting; managing and storing seeds for the next crop; and threshing and milling (CGIAR Gender Programme 1995). On the contrary, women were observed to participate in fewer agricultural activities in wheat farming. Caste also plays an important role in determining women's involvement in agriculture. In a study conducted in the Faizabad district in Eastern Uttar Pradesh, it was found that the women in middle and lower castes participated in agricultural operations in their own rice fields and frequently worked as hired labourers on other's agricultural land. In another study, Paris et al. 2000, found that women from lower castes contributed 60–80% of the total labour input and were involved in almost all operations excluding land preparation and the application of chemicals. Upper caste women were not observed to engage in fieldwork (Harrington et al. 1993). However, the recent large-scale men outmigration has forced many of these women to work in their own fields (FAO 2004). This has not only challenged many pre-existing social notions but has also increased their burden of work.

3.2 Data Source and Methodology

The 2015–16 National Family Health Survey (NFHS-4) data has been used here. The fourth in the NFHS series, NFHS-4 provides information on population, health, and nutrition for India and each state and union territory. The primary objective of the 2015–16 National Family Health Survey is to provide essential data on health and family welfare, as well as data on emerging health issues in India. To achieve a representative sample of 15% of households, NFHS-4 conducted interviews in every alternate selected household in 30% of the selected clusters. In all, 28,586 Primary Sampling Units (PSUs) were selected across the country in NFHS-4, of which fieldwork was completed in 28,522 clusters (IIPS and ICF 2017).

The women data file was used. The available variables from the datasets were used to define the women farmers and the farmers' wives. The women farmers are those who are in agriculture and their husbands are not involved in agriculture. The women farmers work throughout the year or seasonally and in rural areas. The farmers' wives are those women who are in agriculture and their husbands are also in agriculture and the women farmers work throughout the year or seasonally and in rural areas. Based on these definitions, the total number of women farmers and the farmers' wives considered in this study are 3913 and 9323, respectively.

3.3 Results

The understanding about the women farmers and the farmers' wives is important to have a clear vision where these women come from. A brief background description will help us to understand the nuances of the household decision-making.

The household characteristics like caste, religion, and wealth index of the farmers' wives and the women farmers do not differ drastically. Majority are scheduled caste and Hindus. A higher percentage of farmers' wives belong to tribes (14%) than the women farmers (9%). Around two-thirds of these women belong to the poorest and poor categories of wealth index, while 25% belong to the middle category. The mean household size is six which is same across groups. Around 75% of the farmers' wives reported the relation to the household head as 'wife' and 19% reported as daughter or daughter-in-law, while among the women farmers, 13% are head of the household (vs. 3% among the farmers' wives), 61% are wives, 23% are daughter or daughter-in-law.

The farmers' wives (mean age: 35 years) are comparatively older than the women farmers (mean age: 34 years). Among the women farmers, 28% belongs to the age group 40 and above (vs. 36% of the farmers' wives). The level of educational attainment is higher for the women farmers (51% with no education while 29% attained secondary education) than the farmers' wives (56% with no education while 26% attained secondary education). As a reflection of the respondents' characteristics, the husbands of women farmers (mean age: 39 years) are younger than the farmers' wives (mean age: 41 years). The level of educational attainment is higher for husbands of women farmers (24% with no education while 50% attained secondary education) than their counterpart (35% with no education while 41% attained secondary education). Further, it can be seen that 30% of the women farmers reported that their husbands were away from home a month or more months at a time, in the last 12 months while 12% of farmers' wives reported the same. Fifty-eight percent of the women farmers reported that their husbands were away from home 6 months or more months at a time, in the last 12 months, 47% of the farmers' wives reported the same.

3.3.1 Media Habits and Banking

A brief description of the background characteristics enables us to understand where the respondents come from and who they are, and this helps us to draw inferences regarding their activities and habits. For example, if we try to understand the media habits (Table 3.1) of both the categories of women, it reveals that because the women farmers are more educated than their counterpart, one-fourth of them read newspaper, though not daily whereas 86% of the farmers' wives do not read newspaper at all. Women farmers are 5.8 times more likely to read newspaper as compared to the farmers' wives. Majority do not listen to radio while around 47–49% women of both

Table 3.1 Media habits of the farmers' wives and the women farmers, NFHS4 (2015–16)

Media habits	Farmers' wives $N = 9323$	Women farmers $N = 3913$	Total	Adj OR^
Read newspaper or magazine				5.8*
Not at all	86%	81%	84%	
Less than once a week	8%	10%	8%	
At least once a week	4%	6%	5%	
Almost every day	2%	3%	2%	
Listen to radio				0.9
Not at all	90%	89%	90%	
Less than once a week	3%	5%	4%	
At least once a week	4%	4%	4%	
Almost every day	2%	3%	2%	
Watch television				0.9
Not at all	33%	32%	33%	
Less than once a week	9%	9%	9%	
At least once a week	12%	10%	11%	
Almost every day	47%	49%	47%	
Has mobile phone that respondent uses	20%	38%	25%	0.7

Significance level: ***$p < 0.001$, **$p < 0.05$, *$p < 0.01$; ^Covariates adjusted for socio-economic and demographic variables mentioned above. It also includes the other media habits, bank account and access to loan, decision-making authority, economic independence, freedom of mobility, house and land ownership, and conjugal relations. A significant adjusted OR indicates that even after confounding factors have been taken into account, the odds of the 'women famers' are significantly different from the odds of the 'farmers' wives' agreeing with the same statement

Table 3.2 Access to bank accounts and loans of the farmers' wives and the women farmers, NFHS4 (2015–16)

Bank account and access to loan	Farmers' wives $N = 9323$	Women farmers $N = 3913$	Total	Adj OR^
Has bank or savings account that respondent uses	47%	53%	48%	0.4**
Knows programme in their area that give loans to women to start or expand a business	42%	45%	43%	0.6
Ever taken a loan, cash or in kind, from these programmes, to start or expand a business	36%	29%	33%	1.8**

Significance level: ***$p < 0.001$, **$p < 0.05$, *$p < 0.01$; ^Covariates adjusted for socio-economic and demographic variables mentioned above. It also includes the media habits, bank account and access to loan, decision-making authority, economic independence, freedom of mobility, house and land ownership, and conjugal relations. A significant adjusted OR indicates that even after confounding factors have been taken into account, the odds of the 'women famers' are significantly different from the odds of the 'farmers' wives' agreeing with the same statement

the categories watch television daily. Around 38% of the women farmers have mobile phones, while 20% of the farmers' wives have mobile phones.

In addition to the media habits, Table 3.2 depicts that around half of the respondents have bank accounts that they use and more than 40% of the respondents are aware of the programmes in the area that give loans to women to start or expand a business. Thirty-six percent of the farmers' wives took loans, in cash or in kind, from these programmes, to start or expand a business (vs. 29% of the women farmers). The multivariate analysis reveals that the women farmers are 60% less likely to have bank accounts while they are 1.8 times more likely to take loans from the available programmes in their areas to start or expand a business than the farmer' wives.

3.3.2 *Gendered Relations and Negotiated Roles*

The gendered relations are subjective and differ in every household. It cannot be said that women are not empowered because we need to consider the playfulness of intersectionality. For instance, older women may overpower the younger women or vice versa. So, when we talk about decision-making authority of the women in the household, it may have various connotations and mere secondary data will not suffice. However, using the secondary data can be one of the limitations of the paper, but we try to delve deep into the aspect with the limited secondary information we have. Considering the fact that the women farmers are younger, have a higher level of educational attainment and media exposure than the farmers' wives, it can be assumed that the former category of women have higher decision-making than the latter.

The Table 3.3 below reveals that 19% of the women farmers alone can decide on their health care while 12% of the farmers' wives can decide the same. Similarly, the women farmers can decide on large household purchases, visit family or relatives, or decide the use of money earned by their husbands more than their counterpart. Around two-thirds of the women reported that the decisions are taken jointly with their husbands or partners. As predicted, the women farmers stand out to be having higher decision-making authority than their counterpart.

The available variables in the data on economic independence (Table 3.4 and 3.5) are usually around the income earned and its usage. Around two-thirds of the women farmers and farmers' wives decide with their spouses regarding spending their earnings, while 20% of the women farmers and 12% of the farmers' wives decide alone. Husband/partner decide for 18% of the women farmers and 23% of the farmers' wives. The respondents earn either less than their husbands or around the same. Here again, we see as in Table 3.3 on decision-making authority, the women farmers having higher economic independence than the farmers' wives.

In continuation of what we learned from previous sections, Table 3.6 reveals that for going to the market, 63% of the women farmers go alone (vs. 60% among farmers' wives), while 27% go with someone else (vs. 29% among farmers' wives). For going to the health facility, 59% of the women farmers go alone (vs. 55% among farmers' wives), while 35% go with someone else (vs. 37% among farmers' wives). For going to places outside the village, 56% of the women farmers go alone (vs. 54% among farmers' wives), while 36% go with someone else across groups. Going to

Table 3.3 Decision-making authority of the farmers' wives and the women farmers, NFHS4 (2015–16)

Decision-making	Farmers' wives $N = 9323$	Women farmers $N = 3913$	Total	Adj OR^
Person who usually decides on respondent's health care				0.5
Respondent alone	12%	19%	14%	
Respondent and husband/partner	63%	58%	62%	
Husband/partner alone	23%	19%	22%	
Someone else	1%	2%	2%	
Other	1%	2%	1%	
Person who usually decides on large household purchases				0.6
Respondent alone	9%	11%	10%	
Respondent and husband/partner	64%	62%	64%	
Husband/partner alone	23%	21%	22%	
Someone else	3%	5%	3%	
Other	2%	2%	2%	
Person who usually decides on visits to family or relatives				0.9
Respondent alone	9%	11%	9%	
Respondent and husband/partner	67%	63%	66%	
Husband/partner alone	22%	20%	21%	
Someone else	2%	4%	3%	
Other	1%	2%	1%	
Person who usually decides what to do with money husband earns				1.4
Respondent alone	7%	10%	8%	
Respondent and husband/partner	65%	63%	65%	
Husband/partner alone	25%	22%	24%	
Other	2%	3%	2%	
Husband/partner has no earnings	1%	1%	1%	

Significance level: ***$p < 0.001$, **$p < 0.05$, *$p < 0.01$; ^Covariates adjusted for socio-economic and demographic variables mentioned above. It also includes the media habits, bank account and access to loan, other decision-making authority, economic independence, freedom of mobility, house and land ownership, and conjugal relations. A significant adjusted OR indicates that even after confounding factors have been taken into account, the odds of the 'women famers' are significantly different from the odds of the 'farmers' wives' agreeing with the same statement

the market alone is an important factor as women farmers are 5.1 times more likely to go to the market alone as compared to the farmers' wives.

A further understanding of the house and land ownership will help us to draw inferences in a concrete way. So, when it comes to house and land ownership (Table 3.7), 64% of the women farmers do not own a house while 58% of the farmers' wives do not own a house. Ten percent of the women farmers are

Table 3.4 Economic independence of the farmers' wives and the women farmers, NFHS4 (2015–16)

Economic independence	Farmers' wives N = 9323	Women farmers N = 3913	Total	Adj OR^
Respondent alone can decide how to use money	38%	46%	40%	1.3

Significance level: ***$p < 0.001$, **$p < 0.05$, *$p < 0.01$; ^Covariates adjusted for socio-economic and demographic variables mentioned above. It also includes the media habits, bank account and access to loan, decision-making authority, economic independence, freedom of mobility, house and land ownership, and conjugal relations. A significant adjusted OR indicates that even after confounding factors have been taken into account, the odds of the 'women famers' are significantly different from the odds of the 'farmers' wives' agreeing with the same statement

Table 3.5 Economic independence of the farmers' wives and the women farmers, NFHS4 (2015–16)

Economic independence	Farmers' wives *N* = 6630	Women farmers *N* = 2645	Total
Person who usually decides how to spend respondent's cash earnings			
Respondent alone	12%	20%	14%
Respondent and husband/partner	63%	61%	63%
Husband/partner alone	23%	18%	21%
Someone else	2%	2%	2%
Respondent earns cash more than husband/partner			
More than him	18%	17%	18%
Less than him	52%	60%	54%
About the same	26%	18%	24%
Husband/partner does not bring in money	3%	4%	3%
Do not know	1%	1%	1%

self-owners, 15% are joint owners while it is 12% and 17%, respectively, for the farmers' wives. Seventy-five percent of the women farmers do not own land, while it is 67% for their counterpart. Six percent of the women farmers are self-owners, 12% are joint owners while it is 9% and 14%, respectively, for the farmers' wives. It can be concluded that the majority do not own a house or land and if they do, the ownership rights are more with the farmers' wives than the women farmers. This reveals a complex scenario depicting an important role played by intersectionality.

In order to understand the conjugal relations (Table 3.8), it can be seen that higher percentages of women farmers reported that husbands are jealous if they talk to other men than the farmers' wives (34% vs. 28%). Husbands accuse women farmers of unfaithfulness more than the husbands' of the farmers' wives (10% vs. 9%). Higher percentages of women farmers reported that husbands do not permit them to meet female friends (25%), limit contacts with family (20%), insists on knowing where respondent is (23%), do not trust respondent with money (28%) while it is 22%,

Table 3.6 Freedom of mobility of the farmers' wives and the women farmers, NFHS4 (2015–16)

Mobility	Farmers' wives $N = 9323$	Women farmers $N = 3913$	Total	Adj OR^
Usually allowed to go to the market				5.1*
Not at all	11%	10%	11%	
Alone	60%	63%	61%	
With someone else only	29%	27%	28%	
Usually allowed to go to the health facility				0.5
Not at all	8%	7%	8%	
Alone	55%	59%	56%	
With someone else only	37%	35%	37%	
Usually allowed to go to places outside this village				0.5
Not at all	9%	9%	9%	
Alone	54%	56%	55%	
With someone else only	36%	36%	36%	

Significance level: ***$p < 0.001$, **$p < 0.05$, *$p < 0.01$; ^Covariates adjusted for socio-economic and demographic variables mentioned above. It also includes the media habits, bank account and access to loan, decision-making authority, economic independence, freedom of mobility, house and land ownership, and conjugal relations. A significant adjusted OR indicates that even after confounding factors have been taken into account, the odds of the 'women famers' are significantly different from the odds of the 'farmers' wives' agreeing with the same statement

Table 3.7 House and land ownership of the farmers' wives and the women farmers, NFHS4 (2015–16)

House and land ownership	Farmers' wives $N = 9323$	Women farmers $N = 3913$	Total	Adj OR^
Owns a house alone or jointly				1.8
Does not own	58%	64%	60%	
Alone only	12%	10%	12%	
Jointly only	17%	15%	17%	
Alone and jointly both	12%	11%	12%	
Owns land alone or jointly				1.3
Does not own	67%	75%	69%	
Alone only	9%	6%	8%	
Jointly only	14%	12%	13%	
Alone and jointly both	10%	8%	9%	

Significance level: ***$p < 0.001$, **$p < 0.05$, *$p < 0.01$; ^Covariates adjusted for socio-economic and demographic variables mentioned above. It also includes the media habits, bank account and access to loan, decision-making authority, economic independence, freedom of mobility, house and land ownership, and conjugal relations. A significant adjusted OR indicates that even after confounding factors have been taken into account, the odds of the 'women famers' are significantly different from the odds of the 'farmers' wives' agreeing with the same statement

Table 3.8 Conjugal relations of the farmers' wives and the women farmers, NFHS4 (2015–16)

Conjugal relations	Farmers' wives $N = 6698$	Women farmers $N = 2885$	Total	Adj OR^
Husband/partner jealous if respondent talks with other men	28%	34%	30%	1.2
Husband/partner accuses respondent of unfaithfulness	9%	10%	9%	0.8
Husband/partner does not permit respondent to meet female friends	22%	25%	23%	1.1
Husband/partner tries to limit respondent's contact with family	18%	20%	18%	1.1
Husband/partner insists on knowing where respondent is	20%	23%	21%	1.2
Husband/partner doesn't trust respondent with money	24%	28%	25%	0.4**

Significance level: ***$p < 0.001$, **$p < 0.05$, *$p < 0.01$; ^Covariates adjusted for socio-economic and demographic variables mentioned above. It also includes the media habits, bank account and access to loan, decision-making authority, economic independence, freedom of mobility, house and land ownership, and conjugal relations. A significant adjusted OR indicates that even after confounding factors have been taken into account, the odds of the 'women famers' are significantly different from the odds of the 'farmers' wives' agreeing with the same statement

18%, 20%, and 24%, respectively, for the farmers' wives. The women farmers are 60% less likely to be trusted by their husbands in regards to money than the farmers' wives (after controlling the confounding variables).

3.4 Discussion

The study is an attempt to unveil the differences and similarities between the lives of women farmers and the farmers' wives. While understanding 'gender' in agriculture, it is also crucial to understand these two categories of women. Due to the dearth of data, it will be difficult to reveal what percentage of women farmers actually claim to be 'farmers' as usually they are referred to as a 'help'. Though socio-economic factors do not differ, the demographic aspects depict that the women farmers are younger and educated more than the farmers' wives. Consequently, media exposure is higher for women farmers than their counterparts. Despite the fact that the women farmers are younger than the farmers' wives, higher percentages of them are heads of the households and they have out-migrant husbands more than the farmers' wives. A general assumption looking at the two categories of women will be that the women farmers will be more empowered than the farmers' wives. This assumption is strongly supported by demographic factors. Accordingly, it was found that the women farmers have higher decision-making power than the farmers' wives regarding a visit to health care, large household purchases, visit family or relatives, use of money earned by husbands, and spending their earnings. Here a word of caution is needed to understand the way 'decision-making' questions were asked in the field. Whether this decision-making simply indicates nodding of the head by the women after the decision been taken by the household members or raising their voice but unheard so silently succumb to the decision already made or raising the voice and being heard while taking the decision. These three circumstances have three different connotations which generally gets diluted while asking questions on decision-making in the field. In no way, we are questioning the data quality, but these are minute but critical points that highlight the need for gender studies. We can delve deep into the fact by a qualitative study and get the nuances of decision-making process and pattern in the household.

The women farmers have higher mobility in terms of going to the market, health facility, or going outside the village than the farmers' wives. The social norms on women's mobility and most importantly in rural areas to move alone are quite strict. It will be a mistake to conclude just on the basis of the factors from secondary data without understanding the undercurrents. Consequently, it can be seen that the economic independence of women farmers is higher than the farmers' wives. They can not only decide how to spend money earned by them but also by their husbands. However, a complete reverse scenario on ownership of home and land is a matter of deep revelations. The ownership of house and land is higher among farmers' wives than the women farmers. A number of questions arise like whether this difference is because of the age difference between women farmers and farmers' wives or because a higher percentage of husbands of women farmers are out-migrants than the farmers' wives? In rural India, the patriarchy prevails and the women have restricted ownership. This also reflects the challenges pertaining to the women land rights in India.

The conjugal relation of women farmers and farmers' wives reveals that it is better for the latter category than the former except the fact that the women farmers are less trusted by their husbands in terms of money than the farmers' wives. Here again, the question arises whether the socio-demographic factors play any role in determining the conjugal dimensions? Is it because the husbands are out-migrants and that affects the conjugal relations of the women farmers? A deeper dig into the data reveals that the women who are uneducated reported a higher percentage of conjugal differences than those who are either primary, secondary, or higher educated. Further, there is a direct relation with husbands' outmigration and conjugal differences. However, as stated in the aforementioned paragraph, the confounding factors are yet to be recognized to draw a concrete conclusion. The secondary data just touch the tip of the iceberg.

3.5 Conclusion

This study raises several questions which are just the tip of the iceberg. Delving deep in this subject unveils enormous unknown facts that are beyond the purview of our own horizons. 'Gender' is a subject where more than numbers, details matter. There is no denial of the importance of numbers because that depict volume, but it resembles an onion where we peel and unveil layers after layers before we get to the core. Nevertheless, not all women are marginalized or disempowered. The decision-making, mobility, and conjugal relations are determined by the socio-demographic factors and the existing social and cultural norms. At times, we accept and submit to a culture believing it as the truth of life. The study demands a thorough qualitative study to answer some of the questions and question some of the facts.

References

Bardasi E, Wodon Q (2009) Working long hours and having no choice: time poverty in Guinea. Policy Research Working Paper No. 4961, World Bank

CGIAR (1995) Women in rice farming systems: from field to lab and back. CGIAR, Washington, p 1995

Doss CR (2001) Designing agricultural technology for African women farmers: lessons from 25 years of experience. World Dev 29(12):2075–2092

Doss CR, Morris ML (2001) How does gender affect the adoption of agricultural innovations? The case of improved maize technology in Ghana. Agric Econ 25:27–39

FAO (2004) Gender and rice. Food and Agriculture Organization of the United Nations, Rome

Harrington LW, Fujisaka S, Morris ML, Hobbs PR, Sharma HC, Singh RP, Chaudhary MK, Dhiman SD (1993) Wheat and rice in Karnal and Kurukshetra districts, Haryana, India: farmers' practices, problems, and an agenda for action. Haryana Agricultural University, Indian Council of Agricultural Research, International Maize and Wheat Improvement Center, and International Rice Research Institute

International Institute for Population Sciences (IIPS) and ICF (2017) National Family Health Survey (NFHS-4), 2015–16. IIPS, Mumbai

Kabeer N (1999a) The conditions and consequences of choice: reflections on the measurement of women's empowerment. UNRISD Discussion Paper No. 108. http://www.unrisd.org/80256B3C005BCCF9/(httpAuxPages)/31EEF181BEC398A380256B67005B720A/$file/dp108.pdf. Accessed 10 June 2016

Kabeer N (1999b) Resources, agency, achievements: reflections on the measurement of women's empowerment. Dev Chang 30:435–464

Meinzen-Dick R, Johnson N, Quisumbing A, Njuki J, Behrman J, Rubin D, Peterman A, Waitanji E (2011) Gender, assets, and agricultural development programs: a conceptual framework. CAPRi working paper no. 99. International Food Policy Research Institute, Washington, DC

Ngenzebuke RL (2014) The returns of "I do": women decision-making in agriculture and productivity differential in Tanzania. http://www.worldbank.org/content/dam/Worldbank/Feature%20Story/Africa/afr-rama-lionel-ngenzebuke.pdf. Accessed 18 July 2016

Paris T, Singh A, Hossain M, Luis J (2000) Using gender analysis in characterizing and understanding farm-household systems in rainfed lowland rice environments. In: Tuong TP, Kam SP, Wade L, Pandey S, Bauman BAM, Hardy B (eds) Characterising and understanding rain fed environments. International Rice Research Institute, pp 339–370

Quisumbing AR, Pandolfelli L (2010) Promising approaches to address the needs of poor female farmers: resources, constraints, and interventions. World Dev 38(4):581–592

Chapter 4
Women Farmers, Constraints and Policy Around Them to Harness Maximum Benefit

Meena Shekhar and Nirupma Singh

Abstract India is predominately an agriculture based country. A total of 70% of its population is rural, of those households, 60% engaged in agriculture for their main source of income (FAO, India). In developing countries, including India, women are major producers of food. Nearly 70% of women are engaged in it as compared to 63% of all economically active men. About 70% of varied farm work is performed by women and some farm activities like processing and storage are predominantly associated with women. Low self-confidence, lack of knowledge, little access to information and technology, and illiteracy are the major constraints which hinders them to come forward in decision-making activity and significantly limits their potential to take initiative and contribute to production enhancement of their farm field.

Women participation in agriculture will be recognized only when they actively participate as a decision maker. The present paper describes the significance of the contribution of women in agriculture, also describes the constraints in terms of less access to productive resources which disable her to be recognized as active and resource productive member. Women are not recognized as farmers in Indian policies, thereby denying them of institutional supports of the bank, insurance, cooperatives, and government departments (Oxfam, India Blog). These constraints need to be addressed at both policy and implementation levels so that more women can be benefitted to fulfil their dreams to build progressive and productive India. To accelerate the growth and development in agriculture sector, it is essential to bridge the gender gap and empower women with knowledge and technology.

Keywords Participation · Recognition · Constraints · Policy · Gender gap

M. Shekhar
NBPGR, New Delhi, India

N. Singh (✉)
Division of Genetics, Indian Agricultural Research Institute, New Delhi, India

S. Munshi, M. Singh (eds.), *Women Farmers: Unheard Being Heard*, Sustainable Agriculture and Food Security, https://doi.org/10.1007/978-981-19-6978-2_4

4.1 Introduction

Women farmers, backbone of rural and national economies, play a vital role in food production and food security. Various activities of agro-production systems are performed by nearly 60% of women workforce yet their role is microscopic in the society (Oxfam, India Blog 2013). They have little access to resources compared to men. Their entitlements are often neglected, even after the Women Farmer's Entitlement Bill of 2011 is there. However, the Census of India does not define the term 'farmer' and it classifies women as 'cultivators' or 'labourers'. A women cultivator has right of lease or contract over land, whereas a labourer has no such right and works on another person's land for wages. Out of 65% female workers engaged in agriculture, majority are labourers. It is a matter of concern and implies increase in vulnerability of women in agriculture as number of women cultivators has decreased over the years, but the number of women labourers has increased (Swaniti initiative 2020). An estimated 52–75% of Indian women engaged in agriculture are illiterate, an education barrier that prevents them from occupying higher skilled jobs or moving to the non-farm sector (Sanghi et al. 2015). To accelerate the growth in agricultural sector, there is a need to bridge this gap and give equal access to men and women (Munshi 2017).

Paddy is staple food in India. Transplanting, weeding, harvesting, drying, winnowing, and seed storage are the main activity related to production of paddy and all these activities are performed mainly by women. They spend 40.2% of their time per season, performing transplanting (39.1 h), harvesting (29.8 h) and weeding (19.0 h) as the major activities (Compendium 2016). There has been a continuous decline in the proportion of female workers employed in agriculture from 86.1% in 1993–94 to 74.9% in 2011–12 (NSSO Employment—Unemployment Surveys, various years). Post-2010, there has been a decline in the rural female workers in India by 2.7 million. Women are involved in agriculture, from crop selection to land preparation, seed selection, planting, weeding, pest control, harvesting, crop storage, handling, marketing, and processing (Ghosh and Ghosh 2014). Rural women accomplish various livelihood activities like producing agricultural crops, tending animals, processing, and preparing food, working for wages in the rural agriculture sector, collecting fuel and water and caring for family members and maintaining their homes. They have been multitasking.

Women labour in agriculture consists of 43% of total labour in the world agriculture (FAO 2011) and it is necessary to empower women farmers at the grassroots level by providing them knowledge on technical and financial aspects of agriculture with an established identity. There is an urgent need to make communication and information tools which is easily accessible to women (Unnati et al. 2012). The present paper shows that the contribution of women in agriculture is extremely significant. It has been observed that women play a significant role in agricultural development and allied activities. Women face many challenges, ranging from access to resources to land ownership. There is an increasing demand to see changes at policy level for effective designing and implementation of policies from national level to state and to district level which creates space for women farmers to have equal rights toward access of resources, equal wages and improved decision-making in agriculture. All these aspects would bring about change in the existing scenario where women farmers are often left behind in mainstream agriculture.

4.2 Status of Agriculture of India

Nearly half of the workforce of the country is engaged in agriculture sector. It contributes to 17.5% of the GDP (based on prices in 2015–16). India's food grain production has been increasing every year, and the country is among the top producers of several crops such as wheat, rice, pulses, sugarcane and cotton. India is the highest producer of milk and the second highest producer of fruits and vegetables. The agricultural productivity depends on several factors like good quality seeds, land, water and fertilizers, access to agricultural credit and crop insurance, storage and marketing infrastructure, etc. Critically observed, its growth has been fairly volatile over the past decade, ranging from 5.8% in 2005–06 to 0.4% in 2009–10 and −0.2% in 2014–15. Such a variance in agricultural growth has an impact on farm incomes as well as farmers' ability to take credit for investing in their land holdings.

The agriculture sector's contribution to the Gross Domestic Product (GDP) decreased from 54% in 1950–51 to 15.4% in 2015–16, while in other service sector it has increased from 30% to 53% (Statistical Appendix, Economic Survey 2015).

Key issues affecting agricultural productivity are the decreasing size of agricultural land holdings, continued dependence on the monsoon, inadequate access to irrigation, imbalanced use of soil nutrients resulting in loss of fertility of soil, uneven access to modern technology in different parts of the country, lack of access to formal agricultural credit, limited procurement of food grains by government agencies, and failure to provide remunerative prices to farmers. As per the current scenario (Agricultural Situation in India Jan 2021) the Wholesale Price Index (WPI) of pulses, vegetables, oilseeds and paddy increased by 13.04%, 12.24%, 8.29% and 0.68%, respectively in November 2020 as compared to that in November 2019.

4.2.1 GDP Share of Indian Agriculture

The average value of GDP share for India from agriculture was 27.83% with a minimum of 15.41% in 2018 and a maximum of 42.77% in 1967 (Fig. 4.1). The latest value from 2019 is 16.02% in comparison to the world average of 10.19% based on 161 countries data, it is based on the global rankings indicator or use the country comparator to compare trends over time. According to the analysis, the contribution of 'Agriculture, forestry and fishing' (AFF) to India's GDP, refined as Gross Value Add in the segregated set of data, has gone up sharply to almost 18% in

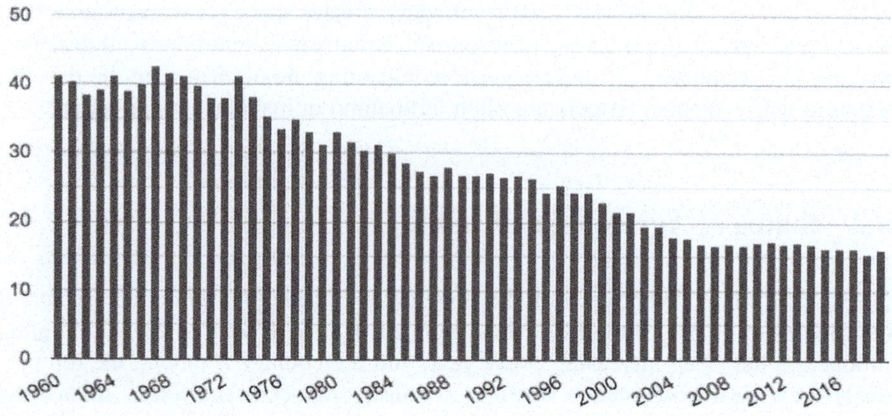

Fig. 4.1 India-GDP share of agriculture

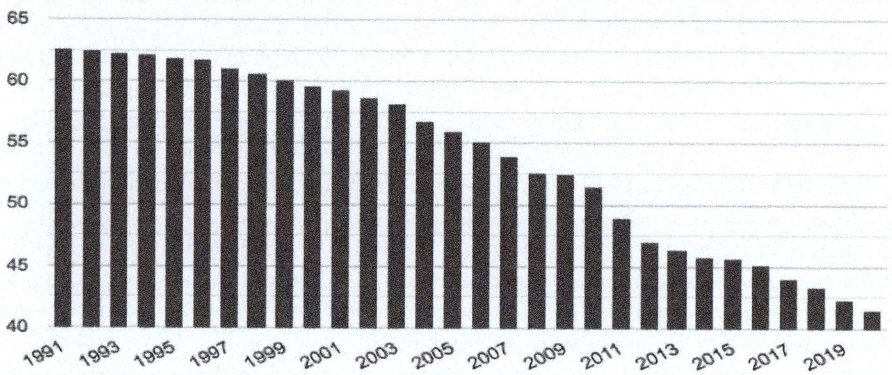

Fig. 4.2 India-Employment in agriculture. Source: Ministry of Statistics and Program Implementation (MOSPI)

the first quarter of 2020–21 from about 13% during April–June quarter of the previous fiscal year. However, the agriculture sector's contribution to national GDP has declined from 34% in 1983–1984 to 16% in 2018–2019.

4.2.2 Employment in Indian Agriculture

Employment in agriculture (Percent of total employment, modelled ILO estimate) in India was reported at 41.49% in 2020. According to the World Bank collection of development indicators, compiled from officially recognized sources. (Fig. 4.2). However, the other half was almost evenly distributed among the two other sectors, industry and services. While the share of Indians working in agriculture is declining,

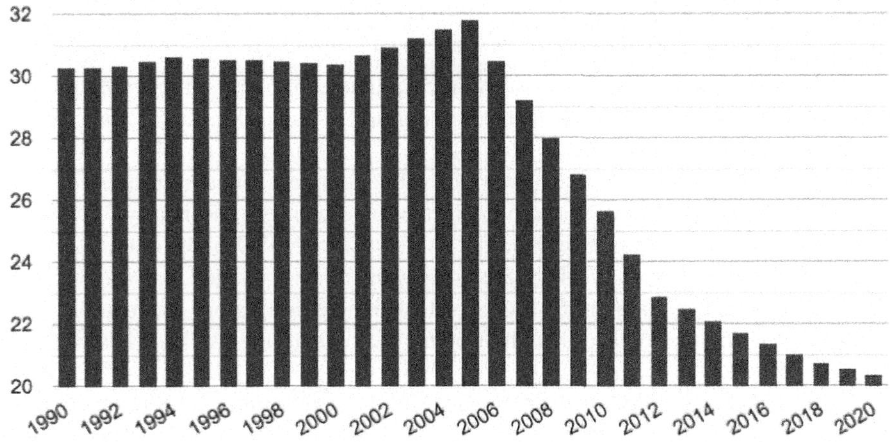

Fig. 4.3 India-Female labor force participation

it is still the main sector of employment. (Percentage of total employment)—actual values, historical data, forecasts and projections were sourced from the World Bank on August of 2021.

In India the average female labour force participation rate from 1990 to 2020 is 27.38% with a minimum of 20.33% in 2020 and a maximum of 31.79% in 2005 (Fig. 4.3). For comparison, the world average in 2020 based on 182 countries is 51.86%.

Over the last three-and-a-half decades, there has been a structural shift in the occupational choice among rural workers, particularly rural agricultural workers, with changes in their occupational choices ranging from agriculture to non-agricultural sectors. According to the 38th Round (1983) of the National Sample Survey (NSS) report, around 77% of rural households depend on the agricultural sector to sustain their livelihoods. Over the years, rural household's dependency on agriculture has declined to 50% as per the latest round of the Periodic Labour Force Survey (PLFS) for 2018–19. In addition, the agriculture sector's contribution to national GDP has declined from 34% in 1983–84 to 16% in 2018–19. Similarly, agriculture sector's contribution to GSDP (Gross State Domestic Product) has broadly followed the same pattern over the same period.

Apart from contributing to the GSDP, the agricultural sector also plays an essential role in generating employment opportunities for the rural workforce. According to the 38th round of NSS and PLFS (2018–19) reports, the agricultural sector's contribution to employment declined from 81% in 1983 to 58% in 2018. This contrasts with rural non-agricultural employment, which increased from 19% to 42% during the same period. In other words, agricultural employment declined by 23%, while employment in the rural non-agricultural sector increased by the same percentage points from 1983 to 2018.

4.2.3 Work Participation of Women in Indian Agriculture

Agriculture is an important sector of the Indian economy, which still holds a majority of the rural workforce for their livelihood. Though the share of agriculture in the national income has fallen from 51.5% in 1950 to less than 14% in 2012–13, almost half of the work force (49%) is still dependent on agriculture for their livelihood. Further, in rural India, 59% of the male workers and nearly 75% of the female workers are engaged in the agriculture sector (NSSO 68th round 2011–12) (Fig. 4.4). Mainly female workers are employed in lower-skilled, lower-paid positions, and are not the owners of capital. Most female cultivators are members of a family that owns the land, rather than being the owners themselves (Kishwar and Vanita 1985).

There has been a continuous decline in the proportion of female workers employed in agriculture from 86.1% in 1993–94 to 74.9% in 2011–12 (NSSO Employment—Unemployment Surveys, various years). Post-2010, there has been a decline in the rural female workers in India by 2.7 million (Fig. 4.5, ICAR-CIAE, Vision Report, 2050, 2014). This change in demographics is slowly changing the face of farming system in India whereas migration of male members of the family has led to what is being called 'feminization of agriculture' in the economic survey 2018. By 2020, women participation is estimated to be around 115 million, that is, 50% in ratio. Though female literacy rate has increased from 8.9% in 1951 to 65.5% in 2011, growth is sluggish in comparison to their male counterparts.

4.3 Status of Women in Indian Agriculture

Women are largely involved in agriculture, more than half of world's food is grown by women, however her hard work has mostly been unpaid all over the world (Dankelman and Davidson 2013). Their indigenous knowledge and skills are very

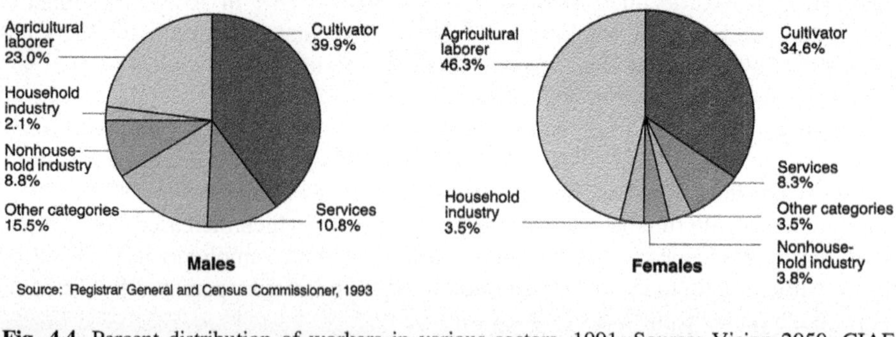

Fig. 4.4 Percent distribution of workers in various sectors, 1991. Source: Vision 2050, CIAE Bhopal

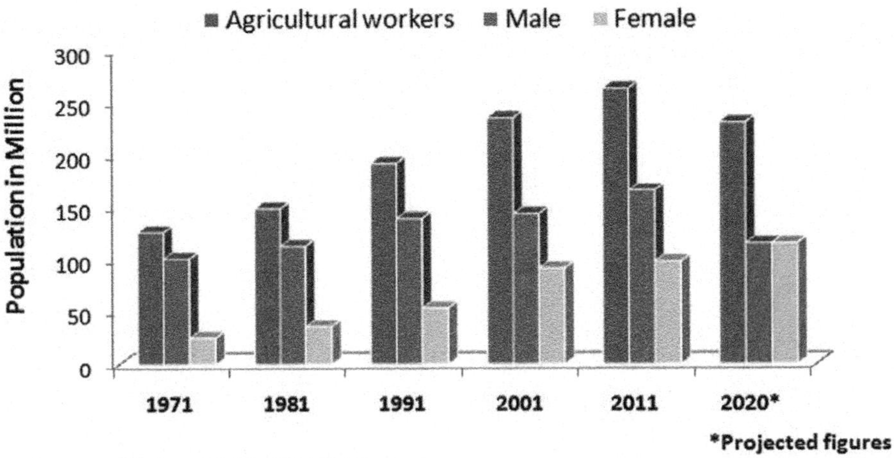

Source: Vision 2050, CIAE Bhopal.

Fig. 4.5 Women dynamics in Indian agriculture

important for food production and sustainable agriculture (Singh and Arora 2010). But often the gender biases act as a barrier in her growth. In a study it was observed that the women had to work harder to receive the benefits as compared to employed man. Women are not perceived as 'farmers' even when they do most of the farm work. This is supported by the fact that there are 5% more female marginal agricultural labourer than the male labourer (Census of India 2011). Agricultural extension and information on new technologies are exclusively directed to men, even though women are increasingly responsible for farm work (Kelkar 2011). Women use conventional tools with little efficiency and face drudgery while working in the field or home (Sethi and Sharma 2011). Women are engaged in the most of agricultural operations from seedbed preparation, sowing, transplanting, weeding, harvesting, and threshing to agro-processing (Gite et al. 2007; Nayak et al. 2013). Women are also employed in other field operations in horticulture, agroforestry, animal husbandry, dairying, and fisheries (Bannerjee 2001). Operations like cleaning, grading, drying, parboiling, milling, grinding, decortication, and storage are also performed by women out of necessity, which is socially accepted. Fodder and fuel collection from forest and farm is another strenuous daily exercise on their part (Ahmad 2011).

It is estimated that women are responsible for 70% of actual work and constitute up to 60% of the farming population reported by Choudhary and Singh (2003). Rural women are responsible for production of more than 55% food grains and comprise 67% of total agricultural labour force. A recent estimate by FAO (2011) showed that women agricultural labour consists of 43% of total labour in the world agriculture. The female small holders access to credit is lower than that of male smallholders (FAO2011), all over the world. In country like India rural women have less land in compare to man and they are less educated than man not because of only external but also due to some internal and personal reasons.

In a country like India where 73.2% of rural women workers are engaged in agriculture, women own only 12.8% of land holdings. In Maharashtra, 88.46% of rural women are employed by agriculture, the highest in the country. In western Maharashtra's Nashik district, women own only 15.6% of the agricultural land holdings, amounting to 14% of the total cultivated area, as per the Agricultural Census of 2015.

4.4 Constraint and Problem Faced by Women

Nearly one half of the population is represented by women who works for two-thirds of the world's workhours. Still they lack control over resources and are poorly represented in positions of power. Such inequalities are noticed all over the world. In a study on the role of women in agriculture which was divided into two categories, viz., monopolizing and supportive role. It was found that they monopolized their role while managing cattles, collection of fodder and selling of livestock whereas their role was supportive in the use of manure, inputs and fertilizers. Ghosh and Mukhopadhyay (1984) revealed that mainly the economic factors determine the workforce participation rates, though some demographic factors were also playing a marginal role. In study it was noticed that the female participation was more prominent in agriculture and was spread over almost all farm activities except ploughing, marketing and irrigation in Gaddi tribals of Himachal Pradesh (Chawhan and Oberoi 1990). At present situation, the gender bias is a universal phenomenon all over the world, and following are the some of issues, the farm women face in her day today life.

4.4.1 Gender Inequalities

In agriculture, women's contribution is not given due recognition as their role is considered to be that of a 'helper' they are not allowed to participate in any decision-making even though they are equally efficient as their male counterparts. Women have insufficient legal land rights and when some of them they do own land, their fields are often relatively small. Men usually tend to cut off women's access to common lands for household needs, and to exacerbate the patriarchal land rights that underpin many customary land rights systems. The other part of gender discrimination is in credit markets, technology, seeds, awareness, and other services, that makes it more difficult for women farmers to participate in out grower schemes. There is inadequate access to extension services, which creates a knowledge gap that prevents them from benefiting equally from new innovations. Women work for longer hours for mostly low and unpaid jobs., Also women's participation in decision making is not acceptable.

In 2011, FAO stated that the yield of farm can be increased by 20–30% if women had the same access to productive resources as men. There was decline in rural main workers from 183 million to 171 million, a reduction of 11.7 million male and a mere 0.5 million female workers during 1991–2001. As males migrate to urban areas to earn more money by working in non-farm activities while women are challenged with the dual responsibility of carrying out farm and household activities They also face varied issues ranging from meagre wages, long working hours, hazardous work and sexual harassment (WTO 2010). Thus, there are five dimensions to gender inequality in agriculture: land rights, productive resources, unpaid work, employment and decision-making (Sexsmith 2017).

4.4.2 Land Ownership/Accessibility Issues

Access to land and security of land tenure affects female farmer's access to other crucial resources such as credit, technology and extension services. Illiteracy among them leads to lack of awareness about their rights and many a time they are not able to avail the benefits of the existing government schemes related to agriculture. The exclusion of women of land ownership rights and the associated resources further creates challenges. Though there are legal reforms as per as women land rights are concerned in India but the actual implementation is a big challenge. Female farmers are unable to utilize the livelihood assets that come from land ownership. Moreover, insecurity in land tenure system implies that most female farmers are often excluded from modern contract-farming arrangements because they do not have full control over their land; a condition which is needed to guarantee the delivery of reliable flow of produce (Amenyah and Puplampu 2013). They face difficulties in getting credit from banks as they are not the rightful owners of the land. Women tends to utilize the income benefits in nutrition, child education, and welfare of the family, whereas men utilize the income mostly for their personal goods and fulfilling individual desires. Therefore, the gender equality in land rights will be a powerful means of eradicating poverty.

4.4.3 Access to Finance and Agricultural Inputs

Poor access to financing is another major hindrance faced by women in agriculture. There may be various causes for the differences in male and female employment and wage patterns. Females are generally less paid for the same amount of work done as by male This is mainly because of unawareness, less education and work experience than men that reduces their bargaining power. Therefore, they easily accept low wages and irregular working conditions. Women are usually paid less than men even for equivalent jobs and similar levels of education and experience (Ahmed and Maitra 2010; Fontana 2009). As per the 66th report of NSSO, the gap in wages between male and female has continued to be significant. Women have no ownership rights over either crop or livestock, and income from all activities (except income from small poultry) highlighted by National Commission for Women, 2005. Many workers have identified the reasons why women farmers are still not able to access credit easily, being the lack of collateral requirements, high transaction cost, limited education and mobility, socio-cultural impediments and irregularity of employment.

Women farmers are deterred from applying for formal loans because of the complexity of the administrative process, unsuitable loan sizes and credit rates. Moreover, women are not found in farmer clusters. Women farmers are not able to access agricultural inputs such as improved seeds, fertilizers, pesticides, machinery, etc., because of the poor financial situation (Sahel Capital and Advisory 2014). Together, these factors place restrictions on access to input and output market information and have a negative impact on women's productivity.

Women's participation in agricultural training program is often low due to the lack of awareness, societal barriers, and transportation facilities. There is a need to make women play effective role, they should be trained to address the local conditions and use of natural resources in a sustainable manner, so that the economic development can be achieved without degrading the environment (Prasad and Singh 1992). However, the agricultural development programs are planned by men and aimed at men. To adopt the modern agricultural technique with utilization of natural resources in sustainable manner, the related training program for rural women is very important. Women's inadequate contact with extension agents is the reason behind agricultural information gap between female and male farmers. There is a productivity gap between men and women farmers due to different levels of access to productive resources and assets stated by the State of Food and Agriculture (2010–11).

4.4.4 Socio-Economic Factors, Participation in Farm Management

Social and economic conditions of women are often the root factor behind her struggles in agriculture. The patriarchal social structure act a barrier as far as her decision-making is concerned It is well established that women farmers are heavily involved in agriculture in rural India although their level of participation in farm management and decision-making are quite low which is attributed to their age, education, land tenancy, and the wealth status. Based on our social, economic and political indicator the status of women in our country is still low. Women constitute 30.3% of the total number of cultivators and 42.6% of the agricultural labourers in India—leading to what can be called as the feminization of agriculture. In India, rural women work as paid labourers, cultivator doing labour on their own land and

Table 4.1 Share of Farm Women in Agricultural Operations

Activity	Involvement (%)
Land preparation	32
Seed cleaning and sowing	80
Inter cultivation activities	86
Harvesting, reaping, winnowing, drying, cleaning and storage	84

Source: Registrar General of India, New Delhi, 2001

Table 4.2 Types of health hazards faced by farm women

Activities	Health hazards reported (%)
Farm activities	
Transplanting	50
Harvesting	26.5
Post-harvest activities	
Threshing	50
Drying	33
Parboiling	67
Livestock management	
Shed cleaning	47
Fodder collection	23
Milching	27.5

Source: Registrar General of India, New Delhi, 2001

managers of certain aspects of agricultural production by way of labour supervision and the participation in post harvest operations. Rural India is witnessing 'feminization of agriculture'. Table 4.1 showed the share of farm women in different agricultural operations. While doing these farm operations women has to face various health hazards. The National Research Centre for Women in Agriculture (NRCW A) has assessed occupational health hazards of farm women in coastal Orissa some years back and results are furnished in Table 4.2.

In rural areas, most of the jobs done by women are very exhausting, moreover they are busy in field work and their girl child looks after home and siblings and leave no opportunity for them to attend school. The dropout rate of girls is higher and their situation remains vulnerable in these circumstances. Increasing food insecurity also has a gender dimension. In a study Udry (1996) and Williams (2004) stated that the shift to cash crops export has also reduced women's bargaining power as their role of food producers is replaced by contingent and seasonal employment and by unpaid family work in men's crops. In addition to agricultural activities women workers are engaged in home activities that limits their exposure to economic opportunities.

4.5 Policy around Women to Harness Maximum Benefits in Agriculture

The challenges faced by women farmers are land rights, decision-making, employment (gender inequalities), product resources, household and multidimensional task, lack of education and unawareness towards agriculture production technologies. To increase the growth in India's agricultural sector, there is a need to bridge gender gap and give equal access to men and women (Munshi 2017). Based on the present status of women in agriculture and their limited access to research and extension services, it is urgently needed to revive the entire approach towards women farmers to cater to their existing needs emphasizing the future challenges that the country has to face in the future. Therefore, empowerment of rural women is the need of the present scenario. Making women as partners in the scenes and projects to bringing them into the mainstream of the agricultural development.

4.5.1 Framework of Strategy to Be Adopted for Women Farmer to Gain Maximum Benefit in Agriculture

A good understanding about the issues and constraint faced by women may be addressed by adopting various policies around women creating opportunities for her to harness maximum benefit. Inclusion of women at the policy designing level can add value New projects development focusing on women farmer. It should be designed the help of various groups and networking as per the local needs and it is to be ensured that the organizational capacity and resources should give more opportunity to women farmer to come forward to participate in various capacity.

- New programs in respect to post-harvest practices, should be initiated by providing crucial back-up services and support to help women groups so that they can adopt new techniques, crops, and enterprises to increase their incomes and employment opportunities.
- Planning for new program with adequate resources needs to be initiated for mobilizing women, creating groups, improving capacity and capability in technical, organizational and commercial (business/micro-enterprises) sectors, and support systems (credit, raw materials, and markets).
- Creation of various new groups jointly in consultations with other organizations (public, private, voluntary) that can potentially complement and supplement the efforts of the public sector (e.g., Ministry or Department of Agriculture).
- Women dependency in agriculture in India is nearly 30.3% of the total cultivators and 42.6% of the agricultural labourers are women in India, leading to the feminization of agriculture. As per *Niti Aayog,* Govt of India and the Research and Information Systems for Developing Countries (2015), increasing women's

ownership of land in rural areas can help in reducing poverty and boost agricultural output in a country.
- The Draft National Policy for Women 2016 reiterates its commitment to achieve Sustainable Development Goles (SDGs) with emphasis on eliminating poverty, inequality, and violence against women by strengthening of policies for empowerment of rural women. It also provides agriculture-based trainings and skill development for farm and non-farm-based entrepreneurship and right to land ownership and nutrition.

4.5.2 Efficient Implementation of the Special Provisions for Women'S Empowerment in Govt Policies in Agriculture

There is a need for efficient implementation of the programs focusing on empowering women in agriculture. A strong monitoring learning and evaluation of each component is required to understanding the gap and strengthening of the deliverables.

- The Government of India has a National Policy for Empowerment of Women (2001) with special emphasis on women in the agriculture and allied sectors.
- National Gender Resource Centre in Agriculture (NGRCA) was set up in the Department of Agriculture, Cooperation & Farmers Welfare (DAC&FW) in 2005–06 to act as a focal point for convergence of all gender-related activities. NGRCA undertakes and supports training, research, and advocacy of mainstream gender issues in agriculture and management of natural resources. Some of its current research activities include the following points:
  (a) Adoption of gender-friendly tools by women farmers and its impact on their lives.
  (b) Schemes for improving women farmer's access to extension services and gender mainstreaming in agriculture.
  (c) Kisan credit card and issues related to credit availability to women.
  (d) Gender-based impact analysis of ATMA (Agriculture Technology Management Agency) program.

- A draft National Policy for Women in the agricultural sector was prepared by The National Commission for Women in 2008, which was followed by the Women Farmers' Entitlement Bill introduced by Prof M S Swaminathan in 2011 as a private member's bill in Parliament. However, this bill was not taken up. There is now a network of over 100 women farmer-organizations across the country, called *Makaam*, which is drafting a revised bill, with support from UN Women and the National Commission for Women on this aspect.

- Women are engaged in agriculture, livestock, fisheries, forestry and allied activities and they need to be recognized as farmers. Their contributions to agriculture and the generation of household incomes have to be recognized and acknowledged. Such recognition will strengthen their legal entitlements and access to inputs and services, with role in household decision-making.
- Women has to face extra burden during peak agricultural seasons and for smooth functioning the specific support services are to be provided for women farmers during this period, especially for child-care.
- There has to be a thrust on developing tools and technologies which are women-friendly to reduce women's work burden, both on-farm and at home. Women friendly technologies are needed for both productive and domestic work, including provision of cooking fuel and clean drinking water, which is accessible. It is to be ensured that women should have sufficient time for these tasks without stretching their working day.
- Closing gender gap under '*Mahila Kisaan Sashatikaran Pariyajana*' by forming self-help group, community institutions of farmers, sustainable agriculture practices etc.

4.6 The Way Forward

Some important changes in the Indian agriculture scenario need to be addressed while formulating strategies for the betterment of women and, through them, the rural economies.

- There is an urgent need to tackle the issues related to sustainable land and water management. Land as a resource is also becoming fragmented in terms of ownership and smaller in terms of operational holdings for agriculture.
- There is increasing 'feminization of agriculture' because of urbanization and migration of male labourers to find employment in the non-agriculture sector (one of the reasons being that agriculture is no longer able to meet family needs). Women are de-facto managing most of the agricultural activities, particularly in drylands, but there is inadequate appreciation of the need to strengthen them, in terms of inputs and knowledge, and providing them farming legal rights.
- Women workers must be given some financial assistance in the form of subsidy for loans. The loan granting procedures and formalities of the bank should be simplified. Interest-free loans should be given to women workers along with possible tax relaxation.
- There is a need to recognize women's labour in agriculture, and it should be accounted in terms of monetary benefits.
- Women should be given priority in accessing credit on soft terms and condition from banks and other financial institutions for setting up their business and for buying properties.

- More facilities should be provided to poor women for procuring agricultural land, livestock and other extension services.
- There must be some separate education policy for women to enhance women's literacy rates.
- To introduce the structural changes in the society, women must be involved in decision-making bodies.
- For women agricultural labour, minimum and equal wages should be fixed by the government and rates of these wages should be reviewed periodically.
- The cultural barriers must be removed by creating awareness among society so that gender discrimination, negative social attitudes towards women is eliminated, and women must be motivated so that they could freely take a better decision and acknowledged as equal partners and contributors to economic growth and development.
- For women, better educational facilities along with vocational courses should be provided for technical knowledge gain. Also, supervisors/advisor should be made available for providing guidance. So that they must be aware regarding their existing rights and legal provisions.
- Regular training programs for women farmers and labours are required time to time to update their knowledge on new agriculture production technology.
- In addition to above, some alternatives employment programs and opportunities should be provided to agricultural women labour in rural areas.

4.7 Conclusion

Rural women are the major contributors in agriculture sector, from crop production, livestock production to cottage industry. Despite such a huge involvement, her role and dignity has yet not been recognized as her status is low by all social, economic, and political indicators. Women spend more time in fetching water, washing clothes, preparing food, and doing agricultural work. Women's wage work is a threat to the male ego and women's participation in multiple home-based economic activities leads to under remuneration for their work. The nature and sphere of women's productivity in the labour market depends on social and economic factors and cannot get on equal terms when compared to men. Their occupational choices are limited due to social and cultural constraints and lack of facilities like transport, and accommodation in the labour sector. Their labour power is considered inferior because of employer's predetermined notion of women's primary role as home-makers. As a result women are considered in the secondary sector of labour market with low pay, low status, and lack of potential upward mobility. Many women in the urban sector work in low paying jobs. Looking at the present status of women in agriculture, it is urgently needed to revamp the entire approach towards women seeing the future food security challenges. Therefore, empowerment of rural women is the need of the hour by making them as 'partner' in the mainstream of agricultural development.

The Justice Verma Committee has emphasized the need for looking at the status of women in its totality. Nearly 40 crore women out of the total of 60 crore female population depend upon crop and animal husbandry, fisheries, forestry, agro processing and agri-business for their livelihood, yet they are handicapped by their inability to obtain equal access to natural resources such as land and water because of male-biased laws. Therefore, it is important to re-examine the current status of women's empowerment in agriculture. Government should make schemes for women and run them successfully with their full involvement. The Energy and Resource Institute (TERI) has provided some practical solutions by focusing on adopting a gender, agriculture, social, efficient natural resource management, health nexus approach. Technological innovation should be gender inclusive, and women collectives must be aligned for impact along multiple SDGs. Lastly, Self Help Groups must be extended as a social enterprise for women so that they can get recognition and protection they deserve. Agricultural education be made gender. Finally, it is concluded that agriculture is central to economic growth where women can learn the best way to grow and cultivate their own nutritious food and sell at markets. So, closing the gender gap in agriculture is imperative if we wish to increase productivity and ensures food security.

References

Agriculture situations in India (March 2021) Wholesale Prices of Certain Important Agricultural Commodities and Animal Husbandry Products at Selected Centres in India 54: Vol. LXXVII

Ahmad N (2011) From policy to projects: rapid gender review/screening tool, MDB-sponsored regional workshops to mainstream gender equality in infrastructure policies & projects Africa regional workshop. Social Development Department, The World Bank, Addis Ababa

Ahmed S, Maitra P (2010) Gender wage discrimination in rural and urban labour markets of Bangladesh. Oxf Dev Stud 38(1):83–112

Amenyah ID, Puplampu KP (2013) Women in agriculture: an assessment of the current state of affairs in Africa. The African Capacity Building Foundation. www.acbf-pact.org

Bannerjee BN (2001) My Great India: economic development & glaring disparities. CBS Publishers & Distributors. Census of India, New Delhi

Chawhan SK, Oberoi RC (1990) Role of tribal women in farm operations. Indian J Home Sci 20(2): 33

Choudhary H, Singh S (2003) Farm women in agriculture operations. Agric Extn Rev 15(1):21–23

Compendium (2016) Training Programme on 'Gender Sensitization for Strengthening Women Perspective in Agriculture' organized at ICAR-CIWA, Bhubaneswar during 28–30 January, 2016

Dankelman I, Davidson J (2013) Women and the environment in the third world alliance for the future. Routledge, p 228. https://doi.org/10.4324/9781315066219. ISBN: 9781315066219

FAO (2011) The state of food and agriculture; women in agriculture closing the gender gap for development. Food and Agriculture Organization of United Nation, Rom. https://www.fao.org/india/fao-in-india/india-at-a-glance/en/

Fontana M (2009) Gender dimensions of rural and agricultural employment: differentiated pathways out of poverty. Paper presented at the FAO-IFAD-ILO Workshop on "Gaps, trends and current research in gender dimensions of agricultural and rural employment: differentiated pathways out of poverty"

Ghosh M, Ghosh A (2014) Analysis of women participation in Indian agriculture. American Research Institute for Policy Development

Ghosh B, Mukhopadhyay SK (1984) Displacement of the female in the Indian labour force. Econ Polit Wkly 19(47):1998–2002

Gite LP, Singh SP, Agarwal N, Majumder J (2007) Women friendly improved farm tools and equipment. Technical bulletin no. CIAE/2007/128. Central Institute of Agricultural Engineering, Bhopal

ICAR-CIAE (2014) Vision 2050. CIAE, Bhopal

Kelkar G (2011) Gender and productivity assets: implications for women's economic security and productivity. Econ Polit Wkly xlvi(23):59–68

Kishwar M, Vanita R (1985) In search of answers: Indian women's voices from Manushi. Zed Books, London

Munshi S (2017) It's time to recognize and empower India's women farmers alumni. Global Shapers Community, IRRI

National Sample Survey Organization 2011–2012 (68th round) https://catalog.ihsn.org//catalog/3281

Nayak J, Singh SP, Moharana G (2013) Occupational health hazard faced by farm women at their workplaces. DRWA (ICAR), Bhubaneswar, pp 1–32

Oxfam India 2013 When women farm India'S land: how to increase ownership? Oxfam India policy brief no. 8. https://www.oxfamindia.org/sites/default/files/Policy%20Brief%20No%208land%20rights.pdf. Accessed 22 May 2017

Prasad C, Singh RP (1992) Farm women: a precious resource. In: Punia RK (ed) Women in agriculture: education, training and development. Northern Book Center, New Delhi

Sahel Capital Partners & Advisory Limited (2014). http://sahelcp.com/wpcontent/uploads/2016/12/Sahel-Newsletter-Volume-7.pdf

Sanghi S, Sirja A, Vijay SS (2015) Decline in rural female labour force participation in India. A relook into causes. Vikalpa 40(3):255–268

Sethi R, Sharma R (2011) Frontier technologies for empowering farm women. Int J Farm Sci 2(1): 142–145

Sexsmith K (2017) Promoting gender equality in foreign agricultural investments: lessons from voluntary sustainability standards. IISD, Winnipeg. https://www.iisd.org/sites/default/files/publications/promoting-genderequality-foreign-agricultural-investments.pdf

Singh S, Arora R (2010) Ergonomic intervention for preventing musculoskeletal disorders among farm women. J Agric Sci 12:61–71

Statistical Appendix, Economic Survey 2015–16, http://unionbudget.nic.in/es2015-16/estat1.pdf

Swaniti initiative (2020). https://www.swaniti.com

Udry C (1996) Gender, agricultural production, and the theory of the household. J Polit Econ 104(5):1010

Unnati A, Ankush GS, Mande AV (2012) Extent of participation of farm women in decision making. J Dairy Foods Home Sci 31(1):72–74

Williams M (2004) Current and emerging gender issues in the multilateral trading system. Background document for the 7th Women's/Gender Affairs Ministerial Meeting (7WAMM) 30 May to 2 June, the Fiji Islands

WTO (2010) World Trade Report 2010 Trade in natural resources; ISBN 978-92-870-3708-4. www.wto.org

Chapter 5
Challenges and Lessons Learned in Mainstreaming Gender into Rice Research and Technology Development: A Case in Eastern Uttar Pradesh, India

Thelma R. Paris

Abstract This chapter presents the challenges and lessons learned in mainstreaming gender into RR4TD in eastern India, in particular. The discussions in this chapter are based on the author's experiences as a social scientist/gender specialist at the Social Sciences Division. The following discussions are divided into the following sections: (1) importance of earlier policy statements in seriously mainstreaming gender in RR4TD; (2) critical challenges in mainstreaming gender in RR4TD; (3) strategies used to overcome these challenges; (4) lessons learned, and finally, (5) the conclusions. The strategies used to overcome these challenges were: (a) built the capacities of agricultural scientists and social scientists of the research teams on farmer participatory approach and gender analysis; (b) conducted socio-economic research including gender analysis; (c) developed a systematic and standardized method of collecting gender-disaggregated information/data; (d) developed a methodology for incorporating social and gender concerns in the stages of participatory varietal selection .The lessons learned were: (a) need to have a deep understanding of gender roles in a given context/situation; (b) political will of research institutions is critical in mainstreaming gender; (c) need to include female agricultural scientists and social scientists in research teams; (d) continue capacity-building programs/activities on how to mainstream gender into RTTD to sustain project activities; (e) social scientists/gender specialists and agricultural scientists should plan and work together; (f) conduct interviews separately for men's and women's groups; (g) involve women in training activities on all aspects of rice production, especially on seed management; (h) RR4TD institutions and local agricultural extension systems need to work with established and reputable Non-Government Organizations and Women's Self-Help Groups

Keywords Mainstreaming gender · Gender analysis · Technology development

---

T. R. Paris (✉)
Former (retired) Socioeconomist-Gender Specialist at the Social Sciences Division, International Rice Research Institute (IRRI), Los Baños, Philippines

S. Munshi, M. Singh (eds.), *Women Farmers: Unheard Being Heard*, Sustainable Agriculture and Food Security, https://doi.org/10.1007/978-981-19-6978-2_5

5.1 Introduction

Within Southeast and South-Asia, rice dominates not only in production and consumption, but it is also inextricably woven into the social and economic fabric of life. Empirical studies reveal that across countries and types of production systems (irrigated, rainfed) and by socioeconomic status, in general, women's labor contribution compared with men in rice farming is higher in South Asia (more than half in Nepal and eastern India) than in Southeast Asia (more than one fourth) on—a per hectare basis (Pandey et al. 2010; Paris et al. 2008a). Women's and men's roles are often conditioned by several interrelated socio-cultural (including class, ethnicity, age, marital status, and religion), economic, and environmental factors. However, gender roles and responsibilities are dynamic and can change over time depending on emerging changes, e.g., climate variability and abiotic stresses, shifts from subsistence to commercialized rice production, increasing mechanization, urbanization, male labor outmigration, technological interventions, and other driving forces (Pandey et al. 2010). For example, the increasing outmigration of men and decreasing male: female ratio in agricultural labor, tend to change the traditional division of labor in rice production, with women not only increasingly providing field labor for rice production but also taking on managerial and decision-making roles on the farm, including choosing rice varieties to be grown (Paris et al. 2009). Yet, women face several constraints in performing these roles because of their lack of access to technical knowledge and technologies which can reduce their drudgery and provide additional income. Agricultural research for development programs has seldom deliberately included women as users and potential beneficiaries.

Poor rice farming households are often faced with several production constraints such as low productivity due to limited land size, stresses such as drought, floods, salinity, and socioeconomic constraints. Rice farming households need rice and rice-related technologies which can help them solve these constraints (Hossain 2006). The International Rice Research Institute (IRRI) is a non-profit agricultural research and training center under the umbrella of the Consultative Group of International Agricultural Research Centers (CGIAR) located in different parts of the world. IRRI's primary goal is to "*improve the health and welfare of rice farmers and consumers, promote environmental sustainability in a world challenged by climate change, and* support *the empowerment of women and youth in the rice industry*" (https://www.irri.org) (CGIAR 1998). Its objectives are to generate and disseminate rice-related knowledge and technology of short and long-term environmental, social and economic benefits to help enhance national rice research systems.

Thus, to improve food production, food security, nutrition, and reduce poverty, the concerns of women farmers and not only those of men should be addressed in rice research for technology development (RR4TD). Gender mainstreaming is a term commonly used in the development community to discuss the process of bringing the concerns and experiences of women and men into development policies and programs for action aimed at achieving gender equality. As such, the needs of women and men can be valued and favored equally (UN Women 2014). Similarly, gender mainstreaming in agricultural research for development (AR4D) is to achieve gender equality. To achieve this goal, gender mainstreaming must be considered in the research and technology development phases, including situational diagnosis, identification of problems, design, testing of proposed interventions, monitoring and evaluation, and impact assessment (Meinzen-Dick et al. 2011). This requires a multi-disciplinary team of scientists, including social scientists (not only economists, to work together to achieve food security, reduce poverty, and gender equality.

This chapter presents the challenges and lessons learned in mainstreaming gender into RR4TD in eastern India, in particular. The discussions in this chapter are based on the author's experiences as a social scientist/gender specialist at the Social Sciences Division. The following discussions are divided into the following sections: (1) importance of earlier policy statements in seriously mainstreaming gender in RR4TD; (2) critical challenges in mainstreaming gender in RR4TD; (3) strategies used to overcome these challenges; (4) lessons learned, and finally, (5) the conclusions.

5.2 Policy Statements

Formal gender policy statements make an explicit commitment to gender research. These are incorporated into the CGIAR's and IRRI's research strategies. Below are the policy statements in the early phases of gender research and sustained through the years.

5.2.1 Gender Research Within the CGIAR

The CGIAR has had a long but varied history of integrating gender analysis into its research portfolio (Meinzen-Dick 2009). Efforts in mainstreaming gender within the CGIAR and IRRI can be traced back to the mid-1980s in the inter-seminar on gender held in Bellagio, Italy. One of the policy recommendations from this inter-center seminar was that:

> For greater awareness of women's roles in agriculture and their special needs as technology users and beneficiaries, international and national agriculture research centers should develop long-term strategies to involve women, where possible, in all phases of research and technology development work (Rockefeller/ISNAR 1985).

To implement this CGIAR policy recommendation, the following suggestions to International Agricultural Research Centers (IARCs) were made (Poats 1990):

> Gender issues must be linked to the entire technology generalization process. Possible areas for consideration of gender-specific concerns in the technology development process should include variety/commodity choice, technology design, crop and livestock management and adoption, and policy research to break current constraints to technology use. Specific considerations of these concerns could take place in farming systems research. IARCS

> should collaborate with national organizations in generating information and methodologies dealing with gender issues. Interdisciplinary teams of scientists should identify specific areas in which gender makes a difference in making IARC work more effective and efficient. Inter-center exchange among natural and social scientists to discuss particular issues in n particular gender in research plans and procedures. National programs should offer more training opportunities for women, find ways to increase the number of female extension workers to reach farm women, and pay specific attention to gender concerns in on-farm research.

CGIAR has a long but varied history of integrating gender analysis into its research priorities. At the systems level, from mid-1991 to 1997, CGIAR had two components: Gender Analysis and Gender Staffing. In 1997, the CGIAR initiated a System-Wide Program (SWI) on Participatory Research and Gender Analysis (PRGA) to further mainstream gender concerns to plant breeding and natural resource management projects. In 2013, several IARCs including IRRI developed Gender Strategies under the CGIAR Gender Platform to strengthen mainstreaming of gender into their respective programs (CGIAR: Gender Platform 2012).

5.2.2 Within IRRI

The Technical Advisory Committee of the CGIAR endorsed IRRI's policy statement in IRRI's strategy (IRRI 1989: 23), stating that:

> IRRI will continue to promote integrating women's concerns into its research projects and joint research with national agricultural research systems (NARS). Specifically, the inclusion of gender analysis explicitly recognizes the contributions of men and women to rice and rice-related activities. Technologies that reduce the burden of rural women without displacing their income-generating capacity will be accorded priority. IRRI will also intensify efforts to recruit qualified women scientists for its research programs and encourage NARS to expand the number of female scientists working in national rice research and extension programs.

What has been stated by IRRI in its strategy statement concerning women was revolutionary concerning unity. In this policy statement, women engaged in rice farming who used to be taken for granted were now acknowledged as users and potential beneficiaries of rice research and technology development. In 2012, gender was further mainstreamed in IRRI under the Global Rice in Science Partnership (GRiSP). A gender strategy was developed to mainstream gender in its overall programs. To support its overall mission and objectives, the goal of GRiSP's gender strategy is to reduce the gender gap in the rice sector. Its specific objectives were: to ensure that the development of GRiSP's products and services along the rice value chain (production, post-harvest, processing) takes gender differences into account, and addresses the specific needs and preferences of women (GRiSP 2013).

5.3 Background of IRRI's Gender Mainstreaming in Eastern Uttar Pradesh, India

5.3.1 IRRI Collaboration with National Agriculture and Extension Systems (NARES) Partners

In November 1988, IRRI and the Extension Department of the Indian Council of Agricultural Research (ICAR) in New Delhi, India, jointly organized an international conference on "Appropriate Agricultural Technologies for Farm Women in India: Future Research Strategy and Linkage with Development Systems." IRRI former Director M.S. Swaminathan initiated this conference to strengthen links between ICAR and IRRI. One of the primary considerations of this conference was to enhance collaboration between the IRRI Women in Rice Farming Systems (WIRFS) program and the Department of Extension at ICAR-Extension Department on gender issues in rice farming. Thus, in 1991, the WIRFS program at IRRI and the Department of Extension at ICAR initiated the first training activities on "Gender Analysis and Its Application in Farming Systems Research Projects," held in West Bengal and Orissa.

In 1992, the Ford Foundation in New Delhi, India, supported IRRI to coordinate collaborative research on "Rainfed Lowland Rice Production Research (RLRP) in Eastern Uttar Pradesh, India." This consortium provided a farming systems research (FSR) perspective on improving rice productivity in stress-prone environments. For several reasons, EUP was chosen as the primary research area for this project. EUP represents the rainfed shallow lowland ecology wherein the yield gap varied from 34.8% in EUP to 59.5% in Assam. Rice farming in EUP is most vulnerable and risk-prone due to complex ecological situations marked by frequent floods or drought, or both. An analysis of the area, production, and yield of rice during the last 10 years shows that the yield is stagnating at around $\leq$2.0 t /ha from 2001–2002 except in the year 2002–2003, 2004–2005, and 2009–2010 due to uneven rainfall distribution which causes excess water stagnation/ drought or both in different years (Diwedi 2012). Aside from these environmental stresses, rice farming households face several economic, social, and cultural constraints. Social and cultural conditions include a rigid caste system, high illiteracy rates, high population, and discrimination against women. Poverty is pervasive compared with the other Indian states due to limited farmlands and low and unstable yields. These constraints and challenges for increasing rice productivity under adverse conditions made eastern India a priority region for rice by IRRI and ICAR.

This RLRP project allowed testing and validating the methodology for incorporating/mainstreaming gender concerns in collaboration with NDUAT. In 1994, IRRI organized an international conference on "Stress Physiology of Rice" held in Lucknow, U.P. India. This highly technical rice conference was traditionally attended mainly by male agricultural scientists, including agricultural economists and socio economists (Singh 1996). The inclusion of social scientists in the technical discussions provided a greater understanding of the social and economic constraints of resource-poor rice farming households in EI. For the first time, gender issues were presented in a technical forum (Paris et al. 1996; Hossain and Laborte 1996).

With IRRI playing a significant role in this initiative, a project on "Participatory Research in Plant Breeding and Gender Analysis" under the CGIAR SWI on PRGA was initiated in selected rainfed lowland villages in eastern India. Two of the research sites in Faizabad, Uttar Pradesh, were the same villages for gender research in collaboration with the Narendra Deva University of Agricultural Technology (NDUAT) in Kumarganj, EUP) (Paris 2000).

5.4 Key Challenges Faced and How They Were Addressed

IRRI's research on unfavorable rainfed lowland and upland rice-based farming systems led to a new approach from the conventional method, "transfer of the technology." In this mode, scientists determine priorities, who generate technology on research stations and laboratories, to be transferred through extension services to farmers. Although this so-called "top-down"and technology-centric approach successfully led to the adoption of rice technologies in favorable irrigated rice areas, the same process did not apply to the adoption of rice technologies in unfavorable lowland and upland environments. Thus a new approach was needed. There was a need for involving social scientists at the initial stage of planning rather than at the end of the project to do an impact assessment. Instead of starting with the knowledge, problems analysis, and priorities of scientists, the new approach begins with the understanding, problem analysis, and importance of farmers and farm families. Instead of the research station as the primary locus of action, experiments are conducted on farmer's fields and with the involvement of the farmer-cultivator. Instead of the scientist as the central experimenter, it is now the farmer, whether a woman or man, and other family members. A pilot project using this approach was agreed to be conducted first in the Narendra Deva Agriculture Technology (NDUAT) in Kumarganj, Uttar Pradesh, in EUP and later was conducted in other states of EI.

However, several challenges were encountered during the planning phase in mainstreaming a gender perspective using the FSR approach. These were: (1) deeply rooted social and cultural beliefs, perceptions, and practices that discriminate against women; (2) lack of female social scientists with skills and interests to do social science research, i.e., gender analysis (3) lack of socioeconomic data, gender-disaggregated data, in particular; and (4) lack of a systematic strategy to include gender analysis in the RR4TD process. The critical challenges in mainstreaming gender in RR4TD and how they were addressed are shown below (Table 5.1):

Table 5.1 Critical challenges in mainstreaming gender in RRTD and how they were addressed

Challenges to gender integration in agricultural research for development	Strategies used to meet these challenges
(a) Social and cultural constraints	(a) Conducted socioeconomic research including gender analysis
(b) Lack of female local social scientists with skills and interests to do social science research, i.e., research on the gender issues	(b) Built the capacities of social scientists to conduct social science research
(c) Lack of comparable gender-disaggregated data/information	(c) Developed a systematic and standardized method of collecting gender-disaggregated data/information
(d) Lack of an accepted process to integrate gender analysis into an ongoing technology development project	(d) Developed a methodology for incorporating social and gender concerns in the stages of PVS

5.4.1 Social and Cultural Constraints

Patriarchal ideology, dowry during the marriage, caste structure, class, and joint/combined families are the social and cultural practices that influence gender roles and gender relations, affecting intrahousehold resource allocation. Women's identity and space intersect and multiply their marginalities where intersectionality is not just about identity—caste, class, and gender but also in terms of space-household, work, community, and society. The predominant force in the social organization of Indian society is patriarchy. Land ownership in India, the acquisition, license, and transfer of property are through the male members of the family. Although the right to inherit property in post-independence India had been assured to female members by law, there are several issues when it comes to the practical implementations on ground. Furthermore, the socialization of girls within the patrilineal form of social organization ensures that women will not be in a position to claim their legal rights (Ghosh 1987). Moreover, the perception that girls will become housewives anyway deprived them of getting formal education and their potential to pursue their professional pursuits. The majority of the women interviewed were illiterate and had never had the experience of being interviewed.

Classification of families by caste is still practiced in EUP.

These deeply rooted socio-cultural beliefs, perceptions, and practices that discriminate against women led to undervaluation of women's unpaid work (family and exchange labor), in the crop, livestock, and homestead activities, lower wages of women compared with men, no or low representation of women in meetings, training activities, and demonstration trials. The women were hesitant to be interviewed, particularly in open spaces and in the presence of men (Paris 2000).

5.4.2 Lack of Female Local Social Scientists with Skills to Do Social Science Research

With the strong support of the Director of Research at NDUAT, Faizabad, plans were laid down to research gender issues. To better understand gender issues within the biophysical and socio-cultural environment, it was essential to have a female local social scientist on the team. The first barrier faced by the team was the unavailability of female social scientists in NDUAT to conduct gender research. During that period, girls enrolled in the Home Sciences Department rather than in the Agriculture Department. However, these changed as more opportunities and incentives were given to female agriculture graduates.

One female professor from the Home Sciences Department agreed to conduct interviews. Unfortunately, after a few visits, she could not continue because she had to bring to and fetch her son from his school. Moreover, it was difficult to synchronize field visits among the research team members. Furthermore, it was not culturally acceptable for a lone female scientist to go with a group of male colleagues on trips to remote villages, even on official field visits.

Another option was to hire an educated local girl from the village who could speak English and collect information. However, due to social restrictions, young girls were not allowed to roam around in the villages. Given the social and cultural limits, wives were more comfortable with female interviewers or with both male and female interviewers and in their private domains, e.g., the kitchen or the homesteads.

5.4.3 Lack of Gender-Disaggregated Data and Information

One constraint of fully integrating gender in rice-based FSR was the lack of gender-disaggregated data. Although agricultural economists collected labor data for costs and returns analysis of rice production, the labor days per hectare contributed by family, exchange, and hired labor were not disaggregated into male and female. Thus, women's unpaid and paid labor contributions were not counted. This led to women's invisibility in statistics and a lack of recognition of their roles as farmers and farm workers. Aggregated data did not provide a clear picture of gender issues and gender relations.

During the earlier phase of the project, social scientists encountered several problems eliciting responses from women. These were: high illiteracy rates among women; no confidence among women to respond to the questions especially when the male spouse is present; difficulty in relying only on the recall method on the labor hours and days spent by the worker for each operation per plot area; unavailability of women for interviews during peak season. The use of structured questionnaires and too detailed and extended formal discussions were not practical to hold the attention of either men or women farmers, particularly during peak cropping periods. Lower caste women were not confident and comfortable being interviewed in the presence of women from the upper caste.

5.4.4 Lack of an Accepted Process to Integrate Gender Analysis into an Ongoing Technology Development Project

It was agreed that an FSR using a participatory approach be adopted. Thus, a major challenge facing managers of institutional breeding programs is to figure out ways to foster increased participation by end-users. Unfortunately, however, "participatory research for development" does not automatically result in the involvement or inclusion of marginalized social groups, including poor women who contribute significantly to rice production, post-harvest, and meal preparation. Moreover, examples of how social, including gender analysis, add an essential dimension to assessing the potential benefits of participatory varietal selection (PVS) are uncommon.

A formal or accepted process was lacking to integrate gender analysis in an ongoing on-FSR project. This meant identifying gender issues and opportunities to consider both men's and women's perspectives in each stage of the research process. This also meant giving women more access to resources and opportunities to reduce the existing gender gaps.

5.5 Strategies Used to Meet these Challenges

Overcoming the challenges mentioned above to mainstream gender into rice-based FSR required transformative change among the research leaders and managers. Below are the strategies and transformations which took place over the years. These strategies were: a) built the capacities of agricultural scientists and social scientists of the research teams on farmer participatory approach and gender analysis; b) conducted socioeconomic research including gender analysis; c) developed a standardized method of collecting gender-disaggregated information/data; and d) developed a methodology for incorporating social and gender concerns in participatory varietal selection (PVS).

5.5.1 *Conducted Socioeconomic Research Including Gender Analysis*

Gender roles differ and vary depending upon the given environment (biophysical, socioeconomic, and cultural). Thus, it is important to include gender analysis in any characterization of research sites. Both qualitative and quantitative methods were used to gain a better understanding of gender roles, gender relations and constraints that men and women face in securing their livelihoods in given rainfed farming systems. Due to social restrictions, group interviews were conducted separately for men and women from different social classes. Information on male and female participation in farm, off-farm and non-farm activities were collected using Participatory Rural Appraisal (PRA) tools. Gender analysis as part of the socioeconomic analysis was used to characterize and understand farm household systems in typical rainfed lowland rice villages in the Faizabad district in eastern Uttar Pradesh. Since then, several socioeconomic studies including gender analysis were conducted as integral in the characterization of several stress-prone environments in EI (Paris et al. 1996, 2000, 2007, 2008a). Researchers were also trained on how to gather, analyze, and report the information from male and female farmers.

The extent of female participation in production in India is determined by a nexus of class/caste hierarchy and norms of patriarchal ideology. The lower castes are officially classified as the backward and scheduled castes. They are considered the most deprived and underprivileged regarding access to resources and social status. Women from the upper castes stay in seclusion or "indoors" and do not engage in manual work to maintain their social status. Women from the lower castes have more freedom to work on their farms and outside their homesteads to earn a living. But aside from working to earn a livelihood for the family, they are not allowed to move outside of the home. This short time and space that the girls go out for wage work expose them to the dynamics of workspaces, e.g., low wage, caste stigma, and hierarchical relationships (Paris et al. 2008b).

Male outmigration on a long-term or short-term period is one of the constraints in increasing rice productivity. In nuclear households, the wife of the male migrant becomes the de facto household head with increased responsibilities on the farm, household, and childcare responsibilities. Female family and exchange labor participation were higher than male family labor among households with migrant members (Paris et al. 2005). Because of the cultural perception that wives do not know anything about rice farming, they are often ignored in training and extension activities. They lack access to technical knowledge, new or improved seeds, and other productive resources. Compared with men, women have lesser access to assets such as land, farm inputs, e.g., improved seeds, water pumps, and machinery (Paris et al. 2013).

5.5.2 Built the Capacities of Female Social Scientists

To address the lack of female social scientists to do gender analysis and fieldwork, the Ford Foundation, which funded FSR research in EI, sponsored three tribal women from Hazaribagh, India, to pursue their M.S. degrees in Social Sciences at the University of Los Banos, Laguna, Philippines. The Ford Foundation also supported several training courses for social scientists and agricultural scientists to enhance their capacities on FSR. It was agreed that gender-focused research should be piloted in two or three rice-growing villages near the experiment station of NDUAT University, Faizabad, eastern Uttar Pradesh.

With the recommendation and the strong support of Dr. RK. Singh, Director of Research of NDUAT, Abha Singh, a sociologist with a Ph.D., was trained and hired to conduct gender research. She received further training at IRRI on gender analysis in RR4TD. She conducted interviews with men and women in the pilot villages in EUP and had shared her findings in publications and conferences. To reduce the gender gap in training programs, particularly dealing with leadership, in 2002, the first Leadership Course for Asian and African Women for Research and Extension (LCAAWRE) was organized in collaboration with IRRI's Training Center. Aside from developing the leadership skills and confidence of Asian and African women in agriculture, the training course aimed to make them effective agents of change in the agriculture sector and trainers of local women on improved crop production, processing, and seed management (Rubzen et al. 2016). This course provided opportunities for female scientists in EUP, India.

5.5.3 Developed a Systematic and Standardized Method of Collecting Gender-Disaggregated Data/Information

A systematic and comparable gender-disaggregated socioeconomic demographic information, labor participation in crop (rice and non-rice), livestock, off-farm and non-farm activities, wages, and participation of the male and female heads of the heads household in critical agriculture-related decisions were gathered in selected research areas in EUP. Labor participation by gender revealed that across stress-prone rice production systems, female labor inputs comprise 74%, 61%, and 50% in sodic, submergence and drought-prone rice environments, respectively of the total labor inputs. These labor inputs came from women from the lower caste households (Paris et al. 2008a). In coastline saline environments in Orissa, India, labor inputs of female family members were higher among the marginal farming households than on small and medium/farms. These results indicate that female family labor increases with poverty (Paris et al. 2007).

Gender-disaggregated data and women's participation in decision-making were incorporated in the baseline socioeconomic surveys under the Stress Tolerant for Rice in Asia and Africa (STRASA) project. These surveys were conducted by the

IRRI Social Sciences Division in eastern India, Nepal, and Bangladesh. Deka and Gauchan (2012) revealed that women's contribution is slightly more than half in selected stress-prone villages in Golaghat, Sibsagar, and Assam, India. Compared with men, women have low sole decision-making power in rice production, marketing, and income-related decision-making activities. In a similar study by Behura et al. 2012, labor participation of women compared with men was higher (in Chattisgarh) at 58% than in Orissa (18%). Results show that the husbands mainly make farming decisions in most cases, particularly in Orissa. However, women are more empowered in making household-related decisions. Thus, labor participation by male and female workers (family and hired) can be compared by the size of land (small, medium, large) by caste, wealth status, production systems (irrigated, rainfed), degree of commercialization, the incidence of male outmigration, and by technology adoption. The impacts of technological change on gender can be assessed.

5.5.4 Developed a Methodology for Incorporating Social and Gender Concerns in the Stages of Participatory Varietal Selection

It is usually fair to say that the earlier use of participation occurs in a breeding process, the more opportunities users are given to influence the objectives, breeding strategy, and final outputs. However, the extent to which users can realize this opportunity depends on the degree of participation.

In 1997, a farmer–participatory plant breeding program for rainfed rice was developed at IRRI in collaboration with the ICAR. The project includes six research sites representing different rice ecosystems in EI. The project is under the umbrella of the CGIAR's System-Wide Initiative on Participatory Research and Gender Analysis (PRGA). The project was carried out in response to the low adoption rates of improved released cultivars in rainfed rice environments. The low adoption rates relate mainly to the inability of the high centralized breeding system to address the enormous diversity of environmental conditions and end-user needs in EI. This project aimed to increase food security by providing varieties capable of producing high and stable yields. The goal of this initiative is to develop, test, and refine methodologies of participatory research and gender analysis as they apply to the development of new technologies in germplasm and natural resource management This project aims to test the hypothesis that farmer participation in rainfed rice breeding can help develop suitable varieties more efficiently. It is also designed to identify the stages in a breeding program where farmer interfacing is optimal. (Farnworth and Jiggins 2003; Courtois et al. 2001).

The project had two components: a) plant breeding component to develop and evaluate a methodology for participatory improvement of rice for heterogeneous environments and to produce breeding materials suiting to farmers' needs; b) social science component (including gender analysis) that aimed to: (1) characterize cropping systems, diversity of varieties grown, and the crop management practices of rice farmers; (2) to analyze male and female farmers' selection criteria and their reactions in a wide to a range of cultivars and breeding lines, and (3) to enhance the capacities of national agricultural research systems (NARS) in participatory research and gender analysis in plant breeding and rice varietal selection (Courtois et al. 2001). The plant breeding components of the project were PPB and PVS, which IRRI plant breeders led.

1. *Participatory Plant Breeding (PPB*). Using the pedigree selection method, farmers and breeders selected individual plants from segregating populations (F4–5) _ of different varietal crosses. Trials were held on-farm and on-station. Plants chosen by breeders and farmers evaluated these genotypes at maturity stages based on grain character and susceptibility to stem borer. The next phase of the research involved field testing and comparing the breeder-selected and farmer-selected materials. The promising lines thus selected were multiplied and supplied to farmers for evaluation and release.
2. *Participatory Varietal Selection.* In PVS, a fixed set of lines (13–25) advance lines and a local check) suited for the specific hydrological conditions in the area were tested on-station and on farmers' fields. The advanced lines were obtained from the IRRI Shuttle Breeding Program and other breeding programs of NARES in eastern India. Two to three farmers per village conducted the on-farm trials under their management level. During pre-harvest and harvest periods, farmers and breeders visually ranked the rice lines grown on-station and on-farm. The Kendall coefficient of concordance and the Spearman's rank correlation coefficient was used to analyze the agreement of ranking of the genotype among farmers, among breeders, and between farmers and breeder (Courtois et al. 2001). PVS includes researcher-managed trials and farmer-managed trials.

   The methodology for mainstreaming socio-cultural and gender analysis in PVS is discussed in this section.

5.5.4.1 Methodology for Mainstreaming Socio-Cultural and Gender Analysis in PVS

There are many ways to ensure that social and gender analyses are considered in PVS. A guidebook was published to build the capacities of rice breeders to incorporate a gender perspective into their projects (Paris et al. 2011). PVS provided an opportunity for social scientists to work closely with plant breeders by helping build more excellent rapport with farmers and including women in each stage of the PVS stages. Below are the steps of PVS, such as setting breeding goals, the evaluation of new rice lines, comprehensive evaluation of new rice lines, the wide diffusion of seeds, and assessment of benefits of PVS (Fig. 5.1).

The methods and tools for integrating social, cultural, and gender analysis in PVS if the varietal selection process are discussed below.

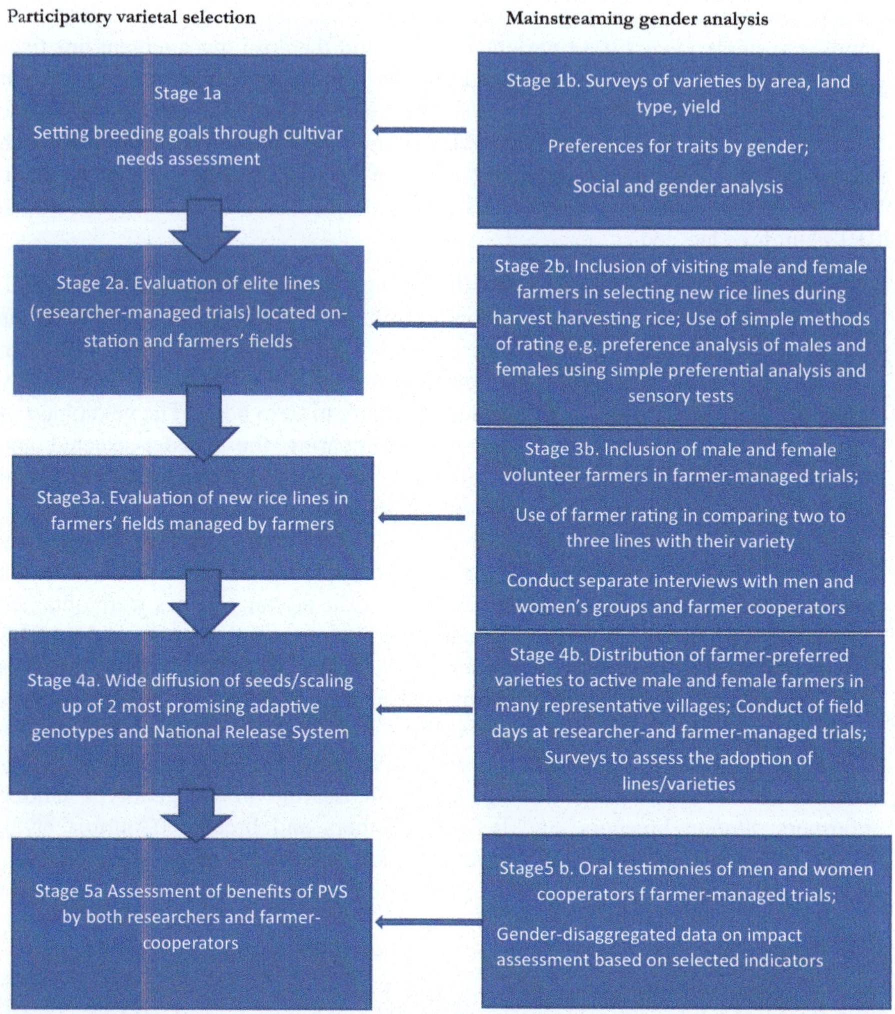

Fig. 5.1 Tools and methods for incorporating socio-cultural and gender analysis in PVS. Source: Paris et al. (2011)

5.5.4.2 PVS and the Integration of Gender Analysis Are Discussed in this Section

Stage 1a: Set Breeding Goals with Farming Communities

In setting breeding goals, the essential steps are: selecting the target site/village which represents the problem of the environment, e.g., drought-prone, submergence-prone, description of the village, understanding when stress occurs, varieties and yields, crop management practices, coping mechanisms, and technology needs and opportunities to increase productivity.

Stage 1b. Gather Information on Socioeconomic Information, and Conduct Gender Analysis, Including Gendered Perceptions of Varietal Preferences

Social research, including culture and gender issues, requires information on the interaction between social class, gender, and labor participation practices. A more gender-responsive agricultural extension, development, and extension system call for a comprehensive look at the same system: who are the actors, the users of the technology, and whose needs are addressed at each stage, from priority setting through the implementation to evaluation and impact assessment.

- Identify gender roles in the household, in on-farm and non-farm activities, and farming practices; quantify the labor contributions of men and women in significant farm activities; assess gender differences in access to and control of resources.
- Identify constraints, potential, and needs; identify options for improvement; Pertinent questions are: Are there gendered differences in rice varietal preferences based on gender roles from rice production, post-harvest, and food preparation? Are there mechanisms to consider the needs of women and men as both producers and consumers? The perceptions that" women do not know anything about rice production and post-harvest traits, etc., automatically excluded them from participating in farmers' meetings, on-farm technology demonstrations, and on-farm trials. As shown in Fig. 5.2, for the first time, the wife was asked what desirable traits they like in a variety.

Fig. 5.2 Identifying gendered perceptions of rice varietal traits

Fig. 5.3 Identifying women's ability to identify rice varieties and performance of rice varieties in stress-prone environments

- Summarize and share the findings with members of the farming communities and research team members. Are there available lines or varieties better than farmers' current varieties? Are farmers willing to test the new lines or varieties in their fields using their management knowledge? As shown in Fig. 5.3, male social scientists can also interview women in the fields where women feel more comfortable expressing their opinions, knowledge, and experience.
- Social scientists play essential roles in eliciting men's and women's knowledge, attitudes, and perceptions on different rice varieties they have grown and would like to grow in the future. Desirable traits should include not only agronomic traits but also cooking, eating, marketable traits, and other post-harvest traits such as being suitable for feeds for the animals, etc.

5.5.4.3 Farmers' Perceptions of Useful Traits in Varietal Adoption

To determine whether there are gender differences in perceptions of useful traits in varietal adoption, the gender research team used graphic illustrations of traits. The team first showed cards that illustrate useful traits in selecting rice varieties They then asked each farmer what traits he or she consider in selecting rice varieties for specific land types—upland and lowland fields. To assess how farmers valued each trait, then this question was asked: "If you have 100 *paise,* how much would you pay for each trait?" The value in *paise* allocated to a particular trait corresponded to the importance given by the farmer. Because many traits are interrelated, the traits were further reclassified in consultation with a plant breeder. Results showed that women farmers are particularly skilled in assessing post-harvest traits such as milling recovery, and the cooking and eating quality of rice. Listening to farmers' perceptions and involving both men and women farmers in selecting rice varieties at the early stage of breeding can lead to faster adoption of varieties suited to their specific rice ecosystems and vice versa (Paris et al. 2001).

Stage 2a. Evaluation of New Rice Lines on Researcher-Managed Located on-Station and Farmers' Fields

A researcher-managed trial is similar to a research station trial, but it can also be conducted in farmers' fields. This includes a) selection of treatments; b) experimental design; c) plot size and plant spacing; d) field layout; e) field operations and data collection. This information is mainly taken by the researchers (plant breeders, agronomists, and field assistants) in collaboration with farmers. Usually, the researchers take charge of crop management and shoulder the costs of the inputs while the farmer lends the plot without remuneration. The researcher and the farmer agree on how the outputs are shared. Researchers include varieties farmers are already using and also new lines. These lines are evaluated during the pre-harvest, harvest, and post-harvest periods.

Stage 2b. Evaluation of New Rice Lines and Farmers' Varieties by Men and Women Farmers during Harvest Season

Plant breeders seldom consult about their rice lines or varieties preferences despite women's active participation in rice production and post-harvest operations. Using simple voting methods, men and women, especially those with no access to formal education, can identify the two most preferred and two most minor select lines in the preferential analysis (Fig. 5.4). At the initial phase of the project, a rule was imposed that at least 30% of the participants should be female to collect reliable information subjected to both qualitative and quantitative analysis of data and information. The preferential analysis and sensory test of cooked rice should be disaggregated by male and female cooperators/participants. These ratings and the information about trial conditions should be recorded in a form that summarize farmers' opinions and preferences. The Kendall coefficient of concordance and the Spearman's rank correlation coefficient can be used to analyze the agreement of ranking of the genotype among farmers (male and female), and between farmers and breeder (Courtois et al. 2001). Separate interviews for male and female farmer-cooperators will encourage women to voice their opinions with confidence.

Fig. 5.4 Women farmers use ballots to vote for preferred rice lines during harvest time

Stage 3: Evaluation of Farmer–Preferred Lines on Farmers' Fields (Farmer-Managed Trials)

Stage 3a: Farmer-managed trials or baby trials are closely associated with farmers' cropping systems Experiments are laid on one side of a farmers' field, and replication is across five areas. Unlike the researcher-managed trials, this usually includes fewer treatments (<3). The steps include: selecting the target site for where farmer-managed trials will be conducted; meeting with farmer-cooperators for farmer-managed PVS; experimental design collection of agronomic data analysis.

Stage 3b: Inclusion of male and female volunteer farmers in farmer-managed trials and use of male and female farmer ratings A farmer-managed trial can be managed by a female (widow *or* de facto head of household) or male farmer. The social scientist conducts baseline surveys of farmer-cooperators of farmer-managed PVS. Farmers grow farmer-selected or preferred lines/varieties on their farms under their management and farm conditions. Men and women are invited to rate the lines using a voting system and coded ballots during harvest season. Farmers compare two to three new lines with their local/traditional variety. After counting the votes, the women and not only the men are invited to express their own opinions on why they selected a specific line or variety. Focus interviews with separate groups (males or females) and individual male and female farmers are also conducted. The farmer-managed trials should contain a group discussion on the performance of the varieties, and farmers should be asked to talk about the good and bad (positive and negative) characteristics of the varieties. These ratings and the information about trial conditions should be recorded in a form that summarizes farmers' opinions and preferences. PVS is mainly organized and led by social scientists to objectively assess which lines/varieties are preferred by farmers rather than just pleasing the plant breeder.

Stage 4a Wide Diffusion of Seeds/Scaling up

The wide diffusion of improved stress-tolerant seeds took more time than expected due to the need for further testing on farmers' fields and approval from the national release system. Moreover, there was a general lack of quality seeds produced at the research station. Farmers with large farm sizes who were able to produce seeds became the local suppliers in the village.

Stage 4b. Distribution of farmer-preferred varieties to active male and female farmers in many representative villages. Seeds of varieties preferred by farmers were disseminated through farmer exchange. The participation of farmers during the field days also encouraged other farmers to test the new lines/varieties. Social scientists began to assess the adoption of specific lines/varieties.

Stage 5a Assessment of Benefits of PVS by both Researchers and Farmer-Cooperators

Based on focus group discussions, the benefits of PVS by both researchers and farmer-cooperators are the following:

- Clearer understanding among plant breeders of farmers' selection criteria. These would be considered in formulating breeding objectives.
- More representation of poor women as visiting farmers in evaluating the performance of new lines in researcher–managed trials.
- Farmers are exposed to many varieties or new lines and have many to choose from.
- Active poor women farmers are included as project cooperators in farmer-managed trials. Both men and women farmer-cooperators can make a more objective evaluation of the new genotypes using their resources.
- Farmers' rights are promoted.
- There is a faster uptake of new varieties in rainfed areas.
- Men and women have better access to improved seeds and new knowledge.
- Varieties are approved from PVS by formal release systems, which consider yields and other traits for poor subsistence-oriented farmers.

During the adoption phase of stress-tolerant seed. An earlier assessment of PVS on women farmers by Paris et al. (2008b) revealed that involvement in PVS and access to new seeds are positively and significantly related to women's empowerment. Women gained confidence in making decisions related to varietal choice, acquisition, and disposal of seeds, and crop management. Participation of both men and women in the early evaluation of the performance of the rice lines/genotypes on their farms led to the development of varieties that are suited to their stress-prone environments. Women's participation in PVS gave them the ability to make choices and increased their access to available seeds, improved gender relations, and appreciation for their knowledge by family members and members of the community.

The Stress Tolerant for Rice in South Asia and Africa (STRASA) project helped women by providing them with flood-tolerant varieties (*Swarna* Sub-1, *Sahbhagi dhan*) and including them in PVS. Including women farmers in PVS also contributes to a better understanding of other traits such as ease of threshing, keeping quality after cooking, and straw quality for animal fodder. For women in flood-prone rice areas, Swarna Sub1-also called "scuba rice" can survive underwater for up to 2 weeks, no longer have to suffer from drudgery, and face health risks since they do not have to replant after severe floods destroy young seedlings. Swarna Sub1, of medium height, is easier to thresh, thus making manual threshing easier for women workers. Moreover, the quality of its rice straw makes for good fodder for their livestock. Thus, they do not have to walk long distances to look for fodder and use their time to take care of their children (GRiSP 2013a).

Early efforts in the dissemination of stress-tolerant seeds gave women access to stress-tolerant seeds, and the opportunity to make decisions as well. as in raising their status in the community. For example, in the Mayurbani district in EI, through the Balasore Social Service Society (BSSS), IRRI introduced and supplied a drought-tolerant variety *Sahbhagi dhan*, after which the Holy Family Catholic Parish started mobilizing the tribal community toward adopting the variety The Odisha Agricultural Development helped provide training and access to facilities. Women led the development of a seed bank, which gave them an important role in seed conservation and the decision-making processes within the family and the community (GRiSP 2013b).

Through focus group discussions (FGDs) conducted with men and women farmers, it became obvious that farmers took pride in being part of the varietal development through PVS trials that allowed them to express their preferences. Furthermore, other traits such as taste and cooking quality need to be assessed carefully. For example, farmers may prefer a bold grain shape or a good volume expansion ratio of cooked rice to give a feeling of satisfaction after eating (Singh et al. 2000).

Mehar et al. (2017) conducted a study on the role of gender, risk, and time preferences in farmers' rice variety selection in eastern India. The results revealed that female farmers who are more risk-averse, usually choose rice varieties based on cooking quality (e.g., good taste, high cooking quality, and good straw quality) and stress tolerance. They are less likely to select hybrid rice and also less likely to have their decision for market-oriented reasons, compared with male farmers.

5.6 Lessons Learned

(a) *Need to have a deep understanding of gender roles in a given context/situation.* Gender roles and responses are variable across and within cultures. Gender roles and gender relations within households are strongly influenced by social (caste, class, ethnicity, religion, etc.), cultural (belief system and practices), economic circumstances (sources of income), family structure (nuclear or extended), stage of women in the life cycle, marital status (single, married), level of education, degree of mechanization, male outmigration, etc. Gender roles are not static and change through time. Therefore, many factors constrain technology adoption, and social scientists play a vital role in problem diagnosis. They should be involved in the initial phases of the technology development process rather than at the impact assessment stage only.

(b) *Political will of research institutions is critical in mainstreaming gender.* Research institutions' written policy statements are essential for implementing programs and plans addressing gender issues. In NDUAT, Kumargani, U.P., the Director of Research, strongly supported the project. Among other things, he provided transportation for the research teams to go to the villages and assigned scientists to the team. IRRI developed strong linkages with ICAR—Center for Women in Agriculture (CWA) through the provision of training, gender audit, and research activities. Gender research was conducted under ongoing RR4TD projects in IRRI and the CGIAR. It is more relevant to integrate gender concerns under the umbrella of ongoing research and development programs that have sufficient budgets for participatory research.

(c) *Need to include female agricultural scientists and social scientists in research teams.* Agricultural scientists (plant breeders, agronomists) and social scientists should listen to women farmers' opinions, assess their attitudes towards certain farms practices, and consider their criteria in the development and dissemination of rice and rice-related technologies. Without a female member in a team, it would be difficult to elicit information from women farmers, particularly in situations where there are social restrictions on women.

(d) *Continue capacity-building programs/activities on how to mainstream gender into RTTD to sustain project activities.* Capacity building for research leaders and managers, agricultural scientists, researchers, mid-level women professionals engaged in research, extension, and development, both men and women farmers should be integral in an institution's plans and programs. For example, in 2002, IRRI Training Center conducted a training program "The IRRI Leadership Course for Asian and African Women for Research and Extension (LCAAWRE) for women professionals in agriculture (Rubzen et al. 2016). A module on integrating a gender is now incorporated in IRRI's Research for Development Training Course for researchers and leaders under the Department of Agriculture in the Philippines.

(e) *Social scientists/gender specialists and agricultural scientists should plan and work together.*

Castillo (1988) remarked: "If one wants to make things happen. One should be in the place where the action is." Ideally, the social scientists and the other research team members (agronomists, plant breeders, social scientists) should visit the field and talk with women farmers." During PVS (harvest time), plant breeders also had the opportunity to interview women and not only men. They were able to elicit first-hand information on the positive and negative traits of the new rice lines. The local social scientists in this project played a pivotal role in bridging the gap between the scientists and the farmers. They provided leadership in organizing meetings/discussions with the people in the village, arranging PVS visits, data analysis, interpretation, and writing jointly authored papers in technical journals and presentations in international and local fora.

Aside from working together in conducting PVS with men and women, farmers, plant breeders, and social scientists co-authored technical publications and presented results in international and national fora (Mackill et al. 2012; Courtois et al. 2001; Paris et al. 2002; Singh et al. 2002)).

(f) *Conduct interviews separately for men's and women's groups*. Village meetings should start with mixed groups followed by single-sex groups to keep men from dominating the discussions. Women can express themselves better if they are with a group of women only. Moreover, the place, time, and duration of meetings for women should be adjusted to accommodate women's time for household and childcare responsibilities. More women will participate in training activities after they have finished their household chores in the morning and when the meeting is conducted within walking distance in the village.

(g) *Involve women in training activities on all aspects of rice production, especially on seed management.* Women's knowledge and skills are vital in making sound decisions on growing rice and post-harvest, particularly when wives become de facto heads of households after husbands migrate for jobs in the cities or abroad. In cases where illiteracy is high among women, visual aids or "hands-on" training or Farmer Field School (FFS) are more relevant than lectures for farmers with lower levels of education Training activities should be conducted separately for men and women.

(h) *RR4TD institutions and local agricultural extension systems need to work with established or reputable Non-Government Organizations (NGOs) and Women's Self-Help Groups.*

During the early phases, there were no Self-Help Groups in the study villages. Thus, it was difficult to mobilize women to participate in project activities. Working with established Self-Help Groups in the village with active women leaders would greatly facilitate the adoption of rice and rice-related technologies. Although there were NGOs in a few villages, these NGOs were more focused on developmental rather than agricultural research for development projects. They have limited skills in conducting research, particularly on gender analysis which requires collecting and analyzing gender-disaggregated data as well as writing reports.

5.7 Conclusions

This chapter has demonstrated how gender issues were integrated into RR4TD conducted by IRRI in collaboration with NARES partners in NDUAT, in Faizabad district, EUP. It discussed the challenges, strategies, and lessons learned.

Despite the many challenges during the early phases of the project, women's inclusion in the rice breeding processes led to several positive outcomes. PVS provided an opportunity to include both men and women in the early evaluation of lines or varieties suitable to farmers' needs in stress-prone rice environments in Faizabad, EUP. Women could express their knowledge and opinions about the new lines/varieties without being intimidated. Plant breeders and agronomists realized that there are gender differences in varietal preferences and women's opinions do matter in the adoption of varieties. The women felt empowered because, for the first time, they were allowed to voice their opinions about their varietal preferences. Participation of both men and women in the early evaluation of the performance of the rice line/genotypes on their farms led to the development of varieties that are suited to their fragile environments. However, involving women in PVS is only the first step toward reducing gender discrimination in project activities and reaching more women. The requirement in PVS to include at least 30% of participants be women among the evaluators, provided opportunities for plant breeders to listen to women's perceptions about their trait preferences. As emphasized by Johnson et al. 2018, simply reaching women does not ensure that they will benefit from a project, and even if women benefit (e.g., increased income or better nutrition), that does not ensure that they will be empowered (e.g., in control over that income or greater participation in decision-making). Although these early efforts failed to measure women's empowerment, the project was able to shift gender morns and attitudes. Reaching women through PVS proved to be a powerful way to increase women's access to information, new seeds, and confidence. However, more women should be provided with training on new knowledge and skills in conserving and managing the seeds, as well as in managing crops and other farm resources.

While there are ongoing efforts to develop and test technologies for women farmers, it is essential to recognize that gender inequalities in access to technologies, labor, and economic returns through gender mainstreaming in RR4TD do not happen automatically. The success of sustaining the inclusion of women depends on the commitment and concerted action of agricultural research systems and policymakers, and women themselves. The greatest challenge for RR4TD institutions is motivating scientists and technologists to undertake to listen and learn: through collaboration with poor women while developing their research priorities and strategies.

Acknowledgments The authors acknowledge Dr. R. K. Singh, former Director of Research of NDUAT, and other NDUAT scientists in Kumarganj, Faizabad, Eastern Uttar Pradesh; Drs. Mahabub Hossain, Virendra P. Singh, Ismael Abdelbagi, IRRI scientists, and the IRRI Gender Research Team namely Dr. HN Singh, Dr. Abha Singh, Amelia Cueno, Joyce Luis, and Donald

Villanueva for their valuable contributions. Special thanks to the men and women farmers and leaders in the research villages.

References

Behura D, Koshta A, Naik D, Samal P, Malabayabas M (2012) Patterns of adoption of improved rice varieties and farm-level impacts in stress-prone rainfed areas and farm-level impacts in stress-prone rainfed areas of Chhattisgarh and Orissa. In: Pandey S, Gauchan D, Malabayabas M, Emerick-Bool R, Hardy B (eds) Patterns of improved rice varieties and farm-level impacts in stress-prone rainfed areas in South Asia. Los Banos, International Rice Research, pp 261–318

Castillo G (1988) Some general impressions and observations from trips to India, Nepal, Bangladesh, Thailand, Indonesia, and the Philippines. In: Filipino women in the Rice farming systems. University of the Philippines and, International Rice Research Institute and Philippine Institute for Development Studies, Diliman, pp 31–50

Consultative Group of International Agricultural Research (CGIAR) (1998). https://www.irri.org

Consultative Group of International Agricultural Research (CGIAR) Research Program on Rice (2013) Global Rice Science Partnership (GRiSP) - gender strategy. International Rice Research Institute (IRRI), Los Banos

Consultative Group of International Agricultural Research (CGIAR): Gender Platform (2012) Strategies for https://gender.cgiar.org/about-us

Courtois B, Singh RK, Bartolome B, Chaudhury D, McLaren G, Misra CG, Mandal NP, Pandey S, Paris T, Piggin C, Prasad K, Roy AT, Sahu RK, Sahu VN, Sarkarung S, Sharma SK, Singh A, Singh HN, Singh ON, Singh NK, Singh RK, Singh S, Sinha PK, BVS S, Takur R (2001) Comparing farmers and breeders ranking in varietal selection for low-input environments: a case of rainfed rice in eastern India. Euphytica 122:537–501

Deka N, Gauchan D (2012) Patterns of adoption of improved rice varieties and farm-level impacts in stress-prone rainfed areas of Assam, India (pp 107–159). In: Pandey S, Gauchan D, Malabayabas M, Emerick Bool M, Hardy B (eds) Patterns of adoption of improved rice varieties and farm-level impacts in stress-prone rainfed areas in South Asia. Los Banos, International Rice Research, pp 105–214

Diwedi JL (2012) Status paper on rice in Uttar Pradesh Available on www.rkmp.co.in

Farnworth CR, Jiggins J (2003) Participatory plant breeding and gender analysis. In: PPB monograph no.4. Participatory Research and Gender Analysis (PRGA) and Consultative Group of International Research (CGIAR)

Ghosh H (1987) Changes in the status of Indian women a case study o Palipur villages. Michigan State University. Working Paper No. 141. East Lansing

GRiSP (Global Rice Science Partnership) (2013) IRRI's Gender Strategy. International Rice Research Institute, Los Baños

GRiSP (Global Rice Science Partnership) (2013a) Women farmers can be choosers. Women in motion. International Rice Research Institute, Los Banos, pp 10–11

GRiSP (Global Rice Science Partnership) (2013b) Tribal women farmers find their voice. Women in motion. International Rice Research Institute, Los Banos, pp 22–23

Hossain M (2006) Rice technology for poverty reduction in unfavorable areas. Paper presented at the Regional Workshop on the Evaluation of IFAD's strategy in Asia and the Pacific. Asian Development Bank (ADB), Manila, Philippines, 26–29 June 2006

Hossain M, Laborte A (1996) Differential growth in rice production in eastern India. Agroecological and socioeconomic constraints. In: Singh VP, Singh RK, Zeigler RS (eds) Proceedings of the international conference on stress physiology of rice, Lucknow, eastern Uttar Pradesh, India, pp 221–239

IRRI (International Rice Research Institute) (1989) IRRI policies. IRRI toward 2020 and beyond. IRRI, Manila, p 2

Johnson N, Balagamwala M, Pinkstaff C, Theis S, Dick MR, Quisumbing A (2018) How do agricultural development projects empower women? Linking strategies with expected outcomes. J Gend Agric Food Secur 3(2):1–19

Mackill DJ, Ismail AM, Singh US, Labios RV, Paris TR (2012) Development and rapid adoption of submergence-tolerant (Sub1) rice varieties. Chapter 6. In: Sparks DI (ed) Advances in agronomy, vol 115. Burlington Academic Press, pp 299–352

Mehar M, Yamano T, Panda A (2017) The role of gender, risk, and time preferences in farmers' rice variety selection in eastern India. Asian J Agric Dev 14:17–36

Meinzen-Dick R (2009) Opportunities and challenges to address gender issues in agricultural development organizations: lesson from a self-Assessment in the CGIAR. Unpublished document. International Food Policy Research Institute

Meinzen-Dick R, Quisumbing A, Behrman J, Jenzano Biermayr P, Wilde V, Noordeloos M, Ragasa C, Beintena N (2011) Engendering agricultural research, development, and extension. In: Research monograph 110. International Food Policy Research Institute (IFPRI), Washington DC

Pandey S, Paris T, Bhandari H (2010) Household income dynamics and changes in gender roles in rice farming. In: Pandey S, Byerlee D, Dawe D, Dobermann A, Mohanty S, Rozelle S, Hardy B (eds) Rice in the global economy: strategic research and policy issues for food security. International Rice Research Institute (IRRI), Los Banos, pp 93–111

Paris TR (2000) Bringing women from the margin to the mainstream of rice research and technology development. Strategies and lessons learned. Unpublished Ph.D. thesis. University of Western Sydney, New South Wales

Paris T, Singh A, Hossain M (1996) Social consequences of stress-prone environment. In: Singh VP, Singh RK, Zeigler RS (eds) Physiology of stress tolerance in rice. Proceedings of the international conference on stress physiology of rice, Lucknow, Eastern Uttar Pradesh, India, pp 207–220

Paris TR, Singh A, Hossain M, Luis J (2000) Using gender analysis in characterizing and understanding rainfed lowland environments. In: Tuong TP, Kam SP, Wade L, Pandey S, Bouman BAM, Hardy B (eds) Characterizing and understanding rainfed environments. Proceedings of the international workshop on characterizing and understanding rainfed environments, 5–9 Dec 1999, Bali, Indonesia. International Rice Research Institute, Los Banos, pp 339–370

Paris TR, Singh A, Luis JS, Singh HN, Singh ON, Singh S, Singh RK, Sarkarung S (2001) Listening to farmers' perceptions through participatory rice varietal selection: a case study in villages in Eastern Uttar Pradesh, India. In: An exchange of experiences from South and Southeast Asia: Proceedings of the international symposium on participatory plant breeding and participatory plant genetic resource enhancement, Pokhara, Nepal, May 1–5, 2000, 459. Centro Internacional de Agricultura Tropical, Cali

Paris TR, Singh RK, Atlin GR, Sarkarung S, Maclaren G, Courtois B, McAllister K, Piggin C, Pandey S, Singh A, Singh HN, Singh ON, Singh S, Singh RK, Mandal NP, Prasad K, Sahu RK, Sahu VN, Sharma ML, Singh RKP, Thakur R, Singh NK, Chaudhury D, Ram S (2002) Farmer participatory breeding and participatory varietal selection in eastern India: lessons learned. In: Whitcombe JR, Parr LB, Atlin GN (eds) Breeding rainfed rice for drought-prone environments integrating conventional and participatory plant breeding in South and Southeast Asia: proceedings of a DFID Plant sciences research programme/IRRI conference, 12–15 march 2002, IRRI Los Banos, Laguna Philippines. Department of International Development (DFID) Plant Sciences Research Programme (PSP), Centre for Arid Zone Studies (CAZS) and International Rice Research Institute (IRRI), Bangor and Manila, pp 57–60

Paris TR, Singh A, Luis J, Hossain M (2005) Labour outmigration, the livelihood of rice farming households, and women left behind. A case study in eastern Uttar Pradesh. Econ Polit Wkly XL (25):2522

Paris TR, Sanjoy S, Singh FP, Mahata KR, Cueno Delos Reyes A, Zolvinski S, Ismail A (2007) Assessing the needs, constraints, and livelihood opportunities in coastal salinity environments: a case in Orissa, India. In: Hoanh CT, Szuster BW, Kam SP, Ismail A, Noble AD (eds) Tropical deltas and coastal zones: food production communities and environment at the land-water interface. CABI Publishing House

Paris TR, Cueno AD, Singh A (2008a) Characterizing and understanding the socioeconomic conditions of farming households in rainfed rice environments: a case in Uttar Pradesh. In improving productivity and livelihood for fragile environments. IRRI Technical Bulletin 2008(13):16–54

Paris TR, Singh A, Cueno AD, Singh VN (2008b) Assessing the impact of participatory research in rice breeding on women farmers: a case study in Eastern Uttar Pradesh, India. Exp Agric 44(1): 97–112

Paris TR, Rubzen Fay Rola, Luis J., Chi T.T., Wongsanum C, and Villanueva D. 2009. Comparative analysis of the impact of labor out migration and remittances on income and rice productivity in the Philippines, Thailand and Vietnam

Paris T, Manzanilla D, Tatlonghari G, Labios R, Cueno A, Villanueva D (2011) Guide to participatory varietal selection for submergence-tolerant rice. International Rice Research Institute, Los Banos

Paris T, Pede V, Luis J, Sharma R, Singh A, Estipular J (2013) Gender caste, and asset control: implications for agricultural projects in rice-wheat systems of eastern India. In: Gender, agriculture and assets: learning from eight agricultural development interventions in Africa and South Asia. Gender and assets project (GAAP). International Food and Policy Institute (IFPRI) and ILRI. GAAP Note November 2013, Washington, DC

Poats S (1990) Gender issues in the CGIAR system: lessons and strategies from within. Paper presented at the 1990 CGIAR Mid-Term Meeting, 21–25 May 1990, the Hague, the Netherlands

Rubzen MF, Paris TR, Luis J, Farivar F (2016) Enhancing women's capacities in agricultural research and development in Asia and Africa. In: Rola-Rubzen MF, Burgess J (eds) Human development and capacity building. Routledge studies in the modern world economy, pp 15–35. Printed by CPI Group (UK) Ltd, Croydon, CRO 4YY

Singh VP (1996) Monitoring and assessing the impact of participatory research for the development of sustainable production systems: a decade experience in rainfed situations of Eastern India. Manila (Philipipnes): International Rice Research Institute.

Singh RK, Singh US, Khush GS, Rohilla R (2000) Genetics and biotechnology of quality traits in aromatic rices. In: Singh RH, Singh US, Khush GS (eds) Aromatic Rices. Oxford & IBH Publishing Co. Pvt. Ltd., New Delhi, pp. 47–70

Singh ON, Singh S, Singh VN, Dwidedi H, Singh A, Singh HN, Singh RK, Paris TR, Atlin AG, and Singh VP. 2002. Addressing genetic improvement and on-farm diversity through farmer participatory breeding on-farm diversity ding: a case study of rainfed rice in Faizabad and Siddharthanagar districts, Eastern Uttar Pradesh, India. In Whitcombe JR, Parr LB, Atlin GN (eds) Breeding rainfed rice for drought-prone environments integrating conventional and participatory plant breeding in south and Southeast Asia: proceedings of a DFID Plant sciences research programme/IRRI conference, 12–15 March 2002, IRRI Los Banos, Laguna Philippines. Department of International Development (DFID) Plant Sciences Research Programme (PSP), Centre for Arid Zone Studies (CAZS) and International Rice Research Institute (IRRI), Bangor and Manila, pp. 57–60

The Rockefeller Foundation/International Service for National Agricultural Research (ISNAR) (1985) Women and agricultural technology: relevance for research, vol 1. Analyses and conclusions. ISNAR, The Hague

UN Women (United Nations) (2014) The World Survey on the role of women-in-development 2014: gender equality and sustainable development. https://www.unwomen.org/en/digital-library/publications/2014/10/world-survey-2014

Chapter 6
Women Farmers in South Asia: Training Needs and Aspiration

Kamrun Naher and Rahat Ara Karim

Abstract In South Asia, women's involvement in the agriculture sector is noticeable than other regions of the world. In this region, women employment in agriculture is relatively higher than men though they are not recognized as the key stakeholders in agriculture. Still, the number of women farmers is increasing, as men particularly young are leaving the farms for better job opportunities in the cities and abroad as well. So, it is likely that future food crop production in this region will largely depend on women farmers. But, are women farmers competent enough to bear the responsibilities? Women's work in agriculture in most South Asian settings is more likely to be due to a distress sale of labor. Women's equality and empowerment in the society as well as in the agriculture sector are being impeded by culture/religion and gender specific constraints like, no/lower ownership of land and other assets, unsuitable working environment in the field, lack of proper knowledge and training, less access to information and extension services, lower access to credit, inputs and marketing of produces, etc. So, to empower the women farmers, all these barriers should be minimized/removed. Besides, their access to modern agricultural technologies should have to be ensured along with intensive training to improve their knowledge and skills in farming activities to feed the future. Women farmers indicated their training need as high in most cases and they need training on homestead gardening, modern techniques of crop production, nursery establishment, livestock and poultry rearing and fish cultivation.

Keywords Equality · Empowerment · Access · Training · Knowledge

---

K. Naher (✉) · R. A. Karim
Food and Agriculture Organization of the United Nations (FAO), Dhaka, Bangladesh
e-mail: kamrun.naher@fao.org

S. Munshi, M. Singh (eds.), *Women Farmers: Unheard Being Heard*, Sustainable Agriculture and Food Security, https://doi.org/10.1007/978-981-19-6978-2_6

Fig. 6.1 Woman farmer working in her field in Beora, Nepal. Source: CDKN (2021)

6.1 Introduction

A large portion of the world's food output originates in the hands of women farmers. In most developing countries, women provide over half the agricultural workforce. Across South Asia, however, women, on an average, account for about 39 percent of the agricultural workforce either as manager or laborers. Country-wise variations range from 50 percent in Bangladesh to only 35 percent in Sri Lanka (Ramachadran 2016). Women play an important role in agriculture sector by involving in crop production activities like land preparation, sowing, weeding, harvesting, livestock rearing, fish production, and selling agricultural products and by products in markets, etc (Figs. 6.1, 6.2, 6.3, 6.4, 6.5 and 6.6). However, their role and status in this sector is varying depending on the culture, age, and region. They face gender-specific constraints like less assets (land, livestock, capital, machineries, etc.), unsuitable environment, lack of training and less access to information, less decision-making power, etc., that lead to less acknowledgment of women contribution of labor in agricultural activities. Moreover, they are underpaid or unpaid for their works. This scenario is more or less common in South Asian countries. Hence, to empower women farmers to feed the world, mainstreaming the gender issues in agricultural policies and extension services, enhance participation of women in agricultural training programs for wide access to information, etc., can be effective.

Fig. 6.2 Women farmers threshing rice in south east Panjab, India. Source: CDKN (2021)

Fig. 6.3 Woman farmer with potato harvest in Himachal Pradesh, India. Source: CDKN (2021)

6.2 Characterization of South Asia

South Asia comprises of eight countries, mainly Bangladesh, Nepal, Sri Lanka, Bhutan, India, Maldives, Afghanistan and Pakistan (Fig. 6.7). It extends south from the main part of the continent to the Indian Ocean. The principal boundaries of South Asia are the Indian Ocean, the Himalayas, and Afghanistan. The Arabian Sea borders Pakistan and India to the west, and the Bay of Bengal borders India and Bangladesh to the east. The western boundary is the desert region where Pakistan shares a border with Iran (WRG 2021). It is one of the most populated region in the world having population around 1.836 billion with 51.63% Male and 48.37% female (World Bank 2019). Hinduism, Islam, and Buddhism are the top three religions of

Fig. 6.4 Weeding at homestead garden in Faridpur, Bangladesh. Source: On-Farm Research Division, Bangladesh Agricultural Research Institute (BARI), Gazipur 1701, Bangladesh

Fig. 6.5 Poultry rearing by woman in Barind, Bangladesh. Source: On-Farm Research Division, Bangladesh Agricultural Research Institute (BARI), Gazipur 1701, Bangladesh

South Asia. Bangladesh, Maldives, Afghanistan, and Pakistan are mainly a Muslim country, while most in India align themselves with Hinduism. In addition, Sikhism is a major religion in the Punjab region, which is located on India's northern border with Pakistan. Hinduism is the main religion in Nepal, but a blend of Buddhism is more prevalent in the north, while Sri Lanka is a country with a strong Buddhist majority and an active Hindu minority.

Fig. 6.6 Woman vaccinating her cow in Pabna, Bangladesh. Source: On-Farm Research Division, Bangladesh Agricultural Research Institute (BARI), Gazipur 1701, Bangladesh

South Asia mainly consists of four climate zones. The northern Indian edge, as well as the upland and mountainous part of Pakistan, has a dry continental, subtropical climate. The far south of India and southwest Sri Lanka are located in the zone of equatorial climates. The rest of the areas lie in the tropical zone, with hot semi-tropical climate in northwest India, cool winter hot tropical climate in Bangladesh, and semiarid tropical climate in the center of the peninsula (Xue and Yanai 2005). The last two tropical climate zones are greatly affected by monsoon circulations (Martyn 1992; Rao 1981; Papadakis 1966).

South Asia is still predominantly an agrarian society, where a majority of the population is dependent on agriculture for their livelihoods. Rural poverty is higher than urban poverty, reflecting the heavy dependence on natural resources that are directly influenced by changes in weather and climate. Adaption in South Asia promotes sustainable agricultural practices and appropriate technologies, innovation to address shorter growing seasons, extreme temperatures, droughts and floods, and strategies for dealing with water shortages, food security and loss of livelihoods (CDKN 2021).

The main characteristics of the major farming systems in South Asia, including the land area and agricultural population as a proportion of the regional total, principal livelihoods and prevalence of poverty are shown in Map 6.1 and Table 6.1 as given below.

Fig. 6.7 Map of South Asia. https://saylordotorg.github.io/text_world-regional-geography-people-places-and-globalization/s12-south-asia.html

6.3 Women Involvement in Agriculture Sector in South Asia[1]

Globally, women's participation in the labor market is changing, which has a positive impact on women empowerment. Table 6.2 illustrates some facts about women employment in different sectors like agriculture, industry, and services in 2019 according to ILOSTAT database. Comparing to different regions of the worlds,

[1]Employment is defined as persons of working age who were engaged in any activity to produce goods or provide services for pay or profit, whether at work during the reference period or not at work due to temporary absence from a job, or to working-time arrangement. The agriculture sector consists of activities in agriculture, hunting, forestry and fishing, in accordance with division 1 (ISIC 2) or categories A-B (ISIC 3) or category A (ISIC 4).

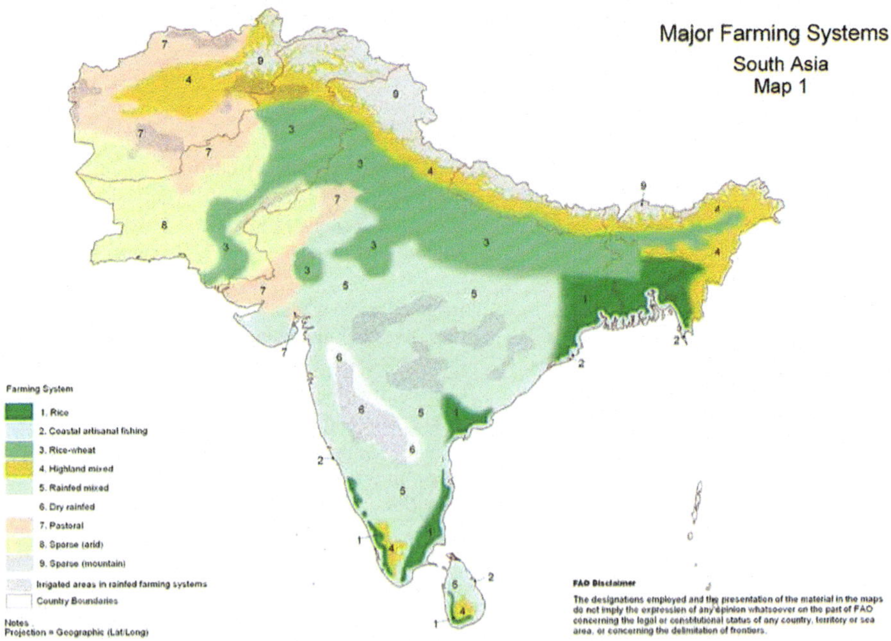

Map 6.1 Major farming systems South Asia. Source: Dixon J. et al. (2001). http://www.fao.org/3/y1860e/y1860e07.htm

the prominence of female employment in agriculture sector in South Asia and Sub-Saharan Africa is striking. Besides in South Asia, the women employment is relatively higher than men in agriculture sector.

Agriculture sector plays an important role in creating women employment opportunities in South Asia. In Pakistan and Nepal, most of the women who work, work in agriculture (Zaidi et al. 2018; Acharya et al. 1999). In Afghanistan, women do as much agricultural work as men, but participate in the more labor-intensive activities in addition to other activities including domestic work (Ashrafi 2009). Agriculture is the chief occupation of women in many parts of Bangladesh. Many ethnic groups such as Santal, Chakma, Garo have been for centuries working as agricultural labor force. The same is also true for many female members of Muslim and Hindus communities (Zaman 2002).

The graph (Fig. 6.8) shows that male and female employment scenario in agriculture sector in South Asian countries in 2019. Overall the engagement of women in agriculture is higher in all countries in South Asia except Maldives. Among all these countries, the percentage of women employment is highest in Nepal in 2019.

In most South Asian rural settings at present, women's agricultural work appears to be driven by household poverty and marginality. Women from wealthier households tend to do less farm work, though with growing male out-migration, they are

Table 6.1 Major farming systems in South Asia (Source: Dixon J. et al. (2001)

Sl. No.	Farming systems	Land area (% of region)	Agric. Popn. (% of region)	Principal livelihoods	Prevalence of poverty[a]
1	Rice	7	17	Wetland rice (both seasons), vegetables, legumes, off-farm activities	Extensive
2	Coastal artisanal fishing	1	2	Fishing, coconuts, rice, legumes, livestock	Moderate—extensive
3	Rice-wheat	19	33	Irrigated Rice, wheat, vegetables, livestock including dairy, off-farm activities	Moderate—extensive
4	Highland mixed	12	7	Cereals, livestock, horticulture, seasonal migration	Moderate—extensive
5	Rainfed mixed	29	30	Cereals, legumes, fodder crops, livestock, off-farm activities	Extensive (severity varies seasonally)
6	Dry Rainfed	4	4	Coarse cereals, irrigated cereals, legumes, off-farm activities	Moderate
7	Pastoral	11	3	Livestock, irrigated cropping, migration	Moderate—extensive (especially) drought induced)
8	Sparse (arid)	11	1	Livestock where seasonal moisture permits	Moderate—extensive (especially) drought induced)
9	Sparse (mountain)	7	0.4	Summer grazing of livestock	Moderate (especially in remote areas)
10	Tree crop	Dispersed	1	Export or agro-industrial crops, cereals, wage labor	Moderate (mainly of agricultural workers)
11	Urban based	<1	1	Horticulture, dairying, poultry, other activities	Moderate

[a] Prevalence of poverty refers to number in poverty, not depth of poverty, and is a relative assessment for this region

taking on farm management roles. Farm laborers are more likely to be from socially marginalized communities, given the negative association of women's farm work with the social standing of a family (Rao et al. 2018). Women's work in agriculture in most South Asian settings, therefore, is more likely to be due to a "distress sale of labor" (Bhalotra 2010) than empowered agency.

Table 6.2 Employment by sector, gender, and region in 2019

	Agriculture		Industry		Services	
	Male	Female	Male	Female	Male	Female
World and different regions	% of male employment	% of female employment	% of male employment	% of female employment	% of male employment	% of female employment
World	27.7	25.3	27.2	15.4	45.1	59.3
East Asia & Pacific	26.9	22.1	29.6	20.4	43.6	57.5
Europe & Central Asia	8.9	6.9	33.8	13.2	57.4	79.9
Latin America & Caribbean	18	7.1	26.3	11.7	55.7	81.3
Middle East & North Africa	14.5	15.8	29.9	14.1	55.6	70.1
North America	1.9	0.8	29.3	8.8	68.8	90.4
South Asia	37.5	57	26.5	17.1	36	26
Sub-Saharan Africa	53.1	52.7	13.3	7.6	33.6	39.7

Source: International Labor Organization, ILOSTAT database. Data retrieved on January 29, 2021. http://wdi.worldbank.org/table/2.3

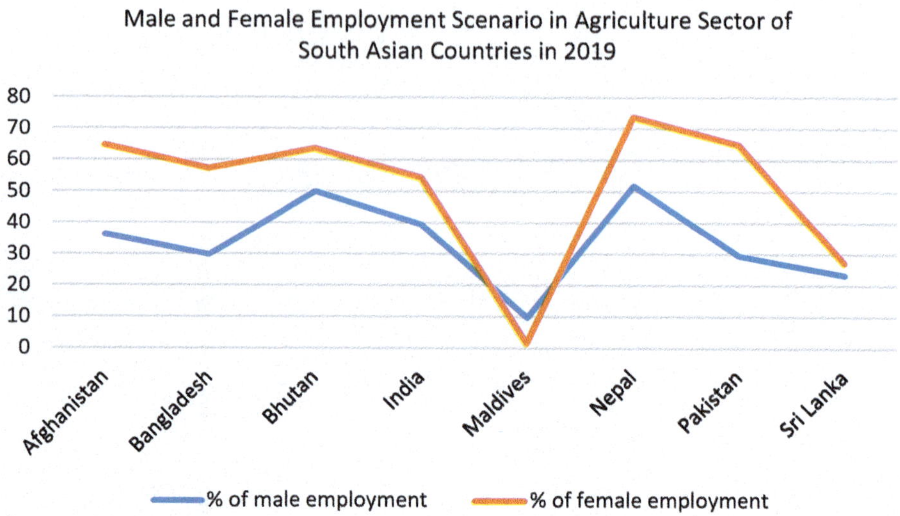

Fig. 6.8 Male and Female Employment Scenario in Agriculture Sector of South Asian Countries in 2019. *Source: International Labour Organization, ILOSTAT database. Data retrieved on January 29, 2021.* http://wdi.worldbank.org/table/2.3

6.4 Constraints as Women Farmers Facing in South Asia

Worldwide, women and girls face unequal opportunities, and Asia is no exception—especially in countries where deep cultural traditions influence the role and mobility of women in society. Improving agriculture production requires access to land, access to knowledge, and access to markets. This is not so easy for women in some Asian countries where men are traditionally or even legally entitled to inherit land or receive a larger share of it as inheritance than women. These customs increase the dependency of women on men, and limit them from owning land which to generate income for themselves and their families (Siddiqi 2015). According to Nguyen et al. (2019), there are some persistent barriers to gender equality in agrarian contexts in South and Southeast Asia. The impediments are as follows:

- Women continue to be regarded as secondary farm labor despite their principal contributions.
- Women's legal access to resources does not always produce control over their use.
- Women's economic participation is confined to the lower levels of the value chain.
- Agricultural policies and implementation sideline gender and power issues.

6.5 Actions Needed to Empower the Women Farmers

To empower women farmers in agriculture sectors, several actions can be taken. The World Bank is now implementing several actions considering the following strategies:

- **Expanding women's access to land and rural finance**: Providing women with greater access to land, finance, and production inputs is critical to closing the productivity gap between men and women. Microfinance institutions and other financial service providers with presence in rural areas can play a key role in supporting women farmers. The Bank also ensures that women benefit from land titling projects.
- **Linking women to agricultural value chains**: When women are linked to agricultural value chains from production all the way to processing and marketing, they help make traditional farming more productive and commercially viable. Inclusive value chains also offer work opportunities for women and men off the farm.
- **Improving rural women's access to training and information**: Knowledge of farming techniques is critical to productivity; however, women farmers have inadequate access to agricultural extension and training services. It is also important that training and agricultural technologies are accessible and adapted to rural women's needs and constraints.

(Source: https://www.worldbank.org/en/topic/agriculture/brief/women-farmers-getting-to-equal)

6.6 Importance of Training for Women Farmers

Knowledge is power, but rural women have little access to education. In times of (economic) crises, girls are the first to be kept at home. Restrictions to the mobility of women due to social norms have a negative effect on her growth and world view. This hampers opportunities for women and girls to access knowledge and available options for trainings, for instance, they also have inhibitions to support innovation on their farms. It also prevents them from learning how to access agriculture advice, such as improving cultivation techniques, land and water management, and pest and disease control. This has a direct effect on how women grow crops, and limits the production levels that could be achieved (Siddiqi 2015). According to Wilson and Hayes (2000), Education and training programs for adults are conducted for five primary purposes:

- To encourage on-going growth and development for individuals
- To assist people in responding to practical problems and issues in adult life
- To prepare people for current and future work opportunities

- To assist organization in achieving desired results and adapting to change and
- To provide opportunities to examine and foster community and societal change

Implicit in each of these five purposes for conducting education and training programs is the expectation of change as an outcome or result (Tennant 2000). Proper knowledge of farming techniques is very important to increase productivity. Improved knowledge on technology and agriculture practices also leads to sustainable change. Since women farmers have inadequate access to agricultural training services, it is very important to train and provide proper information about farming technologies according to their capacity and needs which they can easily adapt. It is extremely important that the designing of the training have to be need based rather aimed to promote one's own invented technology. The women needs, aspirations, and choices have be to be at the core of it .

6.7 Women Farmers' Training Needs

In India women work extensively on farms across the regions—participating in sowing, weeding and harvesting—and are responsible for managing farm work and household chores. However, their contribution in agriculture remains largely unseen and unacknowledged. According to the Food and Agriculture Organization (FAO) of the United Nations, women account for 43 percent of the agricultural labor force in developing countries and produce 60 percent of the food, yet compared with men farmers most women do not have land rights or equal access to education or training. Women farmers recognize that receiving information and skill is more important than short-term monetary support from a project (Dhar 2014).

Yesmin et al. (2007) conducted a study in Netrokona, Bangladesh to assess training needs of rural women. The highest proportion (92%) of the respondents had high training need while about 7% of them had medium training need for income generation activities. They wanted to participate in homestead vegetable production, nursery establishment, livestock and poultry rearing, and cottage and handicraft production but could not be involved due to lack of due knowledge and skill. It indicates that, they felt their training need very much in the above aspects of income generation.

In Rajasthan of India, farm women play an important role in managing different kinds of livestock production operations. However, the degree of their participation in various livestock production activities is governed by the socio-cultural and economic factors. There is poor awareness regarding ways of improving livestock productivity to improve livelihoods. Further, lack of training is also identified as factor working against women farmers' efficiency and effectiveness. Women in some settings are faced with discrimination and have less accessibility to extension services and appropriate technology. Majority of respondent farm women required intensive training on breeding and reproduction management, health management practices, livestock waste management, credit source and credit procurement procedure, insurance, disaster relief and risk management (Dhaka et al. 2017).

According to Jadav et al. (2014), women spent more time than men in dairy production activities. Training is one of the important aspects of empowerment, which can enhance knowledge, improve skill and change the attitude of rural women. Selection of breed, compounding balanced feed using locally available ingredients, feeding during pregnancy, health care and banking and insurance were the most preferred areas of training in dairy farming. The overall adoption level of the trained dairy farm women was higher than untrained dairy farm women.

6.8 Constraints of Women's Participation in Extension and Training

According to Food and Agriculture Organization of the United Nations (FAO 1997), some of the obstacles to women's extremely low access to extension and training are:

- **Lack of knowledge of women's contributions to agricultural production**: Lack of gender-disaggregated data and information, the unpaid nature of much of women's agricultural work, assumptions by agricultural policy makers and planners that agriculture is a male domain, all combine to make women's work "invisible" to field extensionists and extension policy makers.
- **Extension priorities:** Extension services are often directed to cash and export crops and to farmers who own land and have the collateral to obtain credit for inputs and other services. Women farmers are more likely to be responsible for food crops for domestic consumption and less likely than men to own land and have collateral for credit.
- **Agricultural research priorities:** Agricultural research is often directed to cash/export crops and to technologies which are used by male farmers. The food crops raised by women have not generally been a priority in agricultural research. Moreover, there has been little research on technologies appropriate for women, either for their agricultural production or their time-consuming responsibilities for food processing, preparation, and other household tasks.
- **Attitudes and assumptions:** Extension personnel generally share the commonly-held attitudes of society that women do not contribute significantly to agriculture, but are mainly concerned with household responsibilities. It is often assumed that men are the heads of the households and that they will pass on agricultural information to their wives and other women in the household. The fact is that a growing number of rural households are headed by women and that even where men are household heads, they may not transfer information to women, sometimes because it is not relevant to the agricultural work that women are doing.
- **Practical constraints:** Extension services and personnel may not be aware of the practical constraints facing women farmers, such as lack of time due to their household responsibilities in addition to farming; timing of extension services and demonstrations which conflict with women's tasks; restricted mobility for

cultural reasons, lack of money for transport, or inability to leave their children, which may prevent women attending demonstrations or training.

- **Lack of female extensionists:** In many societies, contact between men and women is restricted and, since the great majority of extension workers are male, women farmers may not have access to them.
- **Lack of appropriate extension training materials:** Extension training courses and curricula seldom deal with the role of women in agriculture or approaches for working with women farmers. Many a time the extension materials are gender biased.
- **Low literacy and educational levels of women:** Women farmers are hampered in taking advantage of and using extension services and information because of low literacy rates or lack of access to basic education.

6.9 Interventions Needed to Increase Women's Participation in Training

According to FAO, following interventions can increase the participation of women in training and extension:

- **Data collection and awareness building on women's contributions to agriculture and food security:** The growing collection and dissemination of gender-disaggregated data are contributing to an increasing knowledge and awareness of the important contributions of women to agricultural production and food security. Other measures contributing to this awareness are gender analysis and gender sensitivity training of development policy makers, planners, and agents.
- **Reorientation of extension and research policies and priorities:** Greater knowledge of women's key roles in agriculture can help persuade agricultural development policy makers and planners of the need to reorient extension policies and priorities to include the needs of women food producers and of landless farmers. Mandates and guidelines are needed to implement this, as well as monitoring and evaluation mechanisms.
- **Improving the linkages between extension and research:** Gender-responsive extension services can channel information to research institutes on the needs of women farmers, and gender-responsive research institutes can channel gender appropriate information and technologies to farmers through extension services.
- **Training extensionists to involve women in extension services:** Both men and women extensionists need training on how to work with women farmers and promote their participation in extension work. Some of the efforts being made are training extensionists on gender issues and how to carry out gender analysis; the preparation of specific instructional materials on improving extension work with rural women to be used in special training courses and/or inserted into the curriculum of extension courses and training institutes; and developing training materials appropriate for women.

- **Training women as extensionists:** Girls and women need to be encouraged to train as extension workers. Some efforts in this direction are the provision of special training courses for women farmers, and the reorientation of home economics curricula to emphasize the needs of women in agricultural production.
- **Improving women's access to higher agricultural education and opportunities to benefit from this education:** More women in higher agricultural studies means more potential women extensionists, researchers, and policy makers, and a critical mass of women to help push open the doors of greater opportunity for women.

(FAO 1997)

6.10 Conclusion

Presently women's contribution in agriculture sector is more than men in South Asian countries. As men particularly young are migrating to cities as well as abroad for better job opportunities, responsibility of women in crop production is increasing gradually. But compared to men, women farmers possess poor knowledge and skills in farming activities because of their poor access to training, information and extension services. Social customs and traditions also hinder their empowerment in the society and agriculture sector as well. To achieve women farmers' full potential to feed the future, improving their knowledge and skills in modern agricultural technologies and improved farming activities is indispensable. So, gender-sensitive agricultural policy, gender-responsive extension services, intensive need based training on modern agricultural technologies, easy and affordable access to information and credit, creating linkage between women and agricultural value chain can be effective in this regard.

References

Acharya MPM, Acharya B, Sharma S (1999) Women in Nepal: country briefing paper. Asian Development Bank, Manila

Ashrafi H (2009) Gender dimension of agriculture and rural employment: special focus on Afghan rural women's access to agriculture and rural development sector. Paper presented at the FAO-IFAD-ILO Workshop on Gaps, trends and current research in gender dimensions of agricultural and rural employment: differentiated pathways out of poverty. 31 March–02 April 2009, Rome, Italy

Bhalotra S (2010) Fatal fluctuations? Cyclicality in infant mortality in India. J Dev Econ 93:7–19

CDKN (Climate and Development Knowledge Network) (2021) Agriculture—South Asia. Available at: https://cdkn.org/ar5-toolkit/ar5-south-asia/images-south-asia-ar5/agriculture-images-south-asia-ar5/?loclang=en_gb#:~:text=South%20Asia%20is%20still% 20predominantly,changes% 20in%20weather%20and%20climate

Dhaka BL, Bairwa RK, Meena NL, Meena GS, Chayal K, Nagar BL (2017) Training needs assessment of women farmers on livestock production management in Bundi District of

Rajasthan, India. Int J Curr Microbiol App Sci 6(6):796–803. https://doi.org/10.20546/ijcmas.2017.606.093

Dhar A (2014) Improving women farmer's access to agricultural information and training in India. Available at: https://www.cimmyt.org/news/improving-women-farmers-access-to-agricultural-information-and-training-in-india/

Dixon J, Malcolm H, David GP, Gulliver A (2001) In: Dixon J, Gulliver A, Gibbon D (eds) Farming systems and poverty: improving farmers' livelihoods in a changing world. Malcolm Hall FAO; World Bank, Rome; Washington. http://www.fao.org/3/y1860e/y1860e07.htm

FAO (1997) Higher agricultural education and opportunities in rural development for women-an overview and summary of five case-studies. Information Division, Food and Agriculture Organization of the United Nations (FAO), Rome, Italy

Jadav SJ, DurggaRani V, Mugdal S, Dhamsaniya HB (2014) Women empowerment through training in dairy farming. Asian J Dairy Food Res 33:147–153. https://doi.org/10.5958/0976-0563.2014.00592.2

Martyn D (1992) Climates of the world. developments in atmospheric science, vol 18. Elsevier, Amsterdam

Nguyen HA, Mortensen S, Provalprukskul P (2019) Pathways for Women's empowerment in agriculture in south and Southeast Asia. Stockholm Environment Institute. Available at: https://www.jstor.org/stable/resrep22985?seq=1#metadata_info_tab_contents

Papadakis J (1966) Climates of the world and their agricultural potentialities. Cordoba, Buenos Aires

Ramachadran N (2016) Women and food security in South Asia: current issues and emerging concerns, WIDER research paper, no. 2006/131, ISBN 9291909157, The United Nations University World Institute for Development Economics Research (UNU-WIDER), Helsinki. Available at: https://www.econstor.eu/bitstream/10419/63421/1/52139452X.pdf

Rao N, Gazdar H, Chanchani D, Ibrahim M (2018) Women's agricultural work and nutrition in South Asia: from pathways to a cross-disciplinary, grounded analytical framework. Food Policy 82(2019):50–62. https://doi.org/10.1016/j.foodpol.2018.10.014. https://www.sciencedirect.com/science/article/pii/S0306919218308339

Rao YP (1981) The climate of the Indian subcontinent. In: Takahashi K, Arakawa H (eds) Climates of Southern and Western Asia. Elsevier, Amsterdam, pp 67–117

Siddiqi Y (2015) Women farmer can make Asia more food secure. Available at: https://blogs.adb.org/blog/women-farmers-can-make-asia-more-food-secure#:~:text=This%20varies%20from%20South%20Asia,jobs%20like%20drying%20and%20storage

Tennant M (2000) Adult learning for self-development and changes. In: Wilson AL, Hayers ER (eds) Handbook of adult and continuing education. Jossey-Bass, San Franscisco

WRG (2021) Chapter 9: South Asia. World Regional Geography (WRG): people, places, and globalization. University of the Minnesota Libraries Publishing. Available at: https://open.lib.umn.edu/worldgeography/part/chapter-9-south-asia/

Wilson AL, Hayes ER (2000) Handbook of adult and continuing education. Jossey-Bass, San Francisco

World Bank (2019) Population, total—South Asia. Available at: https://data.worldbank.org/indicator/SP.POP.TOTL?locations=8S

Xue Y, Yanai M (2005) Asia, climate of south. In: Oliver JE (ed) Encyclopedia of world climatology. Encyclopedia of earth sciences series. Springer, Dordrecht. https://doi.org/10.1007/1-4020-3266-8_20

Yesmin S, Rahman MZ, Miah MAM, Akhter S (2007) Training need of rural women participating income generating activities of SUS. Prog Agric 18(2):271–278

Zaidi Y, Farooq S et al (2018) Rural Women in Pakistan – Status Report 2018 UN Women Pakistan. https://www.ncsw.gov.pk/SiteImage/Downloads/Rural%20Women%20in%20Pakistan%20Status%20Report%202018.pdf

Zaman S (2002) Women's role and status in Bangladesh agriculture. Human Development Research Centre (HDRC), Dhaka

Chapter 7
Is Scale-Appropriate Farm Mechanization Gendered? Learning from the Nepal Hills

Hom N. Gartaula, Madhulika Singh, and Gokul P. Paudel

Abstract Scale-appropriate farm mechanization could be an important pathway to the UN's sustainable development goals (SDGs) of attaining gender equality (SDG5) in agriculture. Gender and farm mechanization is getting attention in the academic and public policy domain as a solution to labor scarcity in the smallholder farming systems, which in recent years, is facing challenges of labor shortage due to male labor outmigration. Taking a case study from a maize-based farming system in Nepal hill, this paper illustrates how the promotion of scale-appropriate farm mechanization can be gendered. Using the household survey data collected from the mid-hills of Nepal from 179 mini-tiller adopter farmers, this paper reports that only 4% of the owners were women, and only 1% of women were involved in mini-tiller operations. We find that mini-tiller adopting male and female household head's maize productivity, profitability, and production costs are similar. The paper concludes by identifying social perception against women, rugged topography, women's low level of knowledge in operating machines, spare parts maintenance, and added responsibilities resulting in women's lower participation in mini-tiller adoption decisions. This chapter suggests measures like awareness raising, increasing access, and training built around tailoring women's needs to reduce the gender gap in farm mechanization.

Keywords Mechanization · Gender · Equality · Outmigration · Labor-shortage

---

H. N. Gartaula (✉) · M. Singh
International Maize and Wheat Improvement Center (CIMMYT), New Delhi, India
e-mail: h.gartaula@cgiar.org

G. P. Paudel
International Maize and Wheat Improvement Center (CIMMYT), Lalitpur, Nepal

School of Economics and Management, Leibniz University Hannover, Hannover, Germany

S. Munshi, M. Singh (eds.), *Women Farmers: Unheard Being Heard*, Sustainable Agriculture and Food Security, https://doi.org/10.1007/978-981-19-6978-2_7

7.1 Introduction

Among the 17 Sustainable Development Goals (SDG) of the UN, SDG5 aims to achieve gender equality and empower all women and girls by 2030 (UN 2015). It is challenging in agriculture, where almost 45% of the rural women provide labor forces worldwide (FAO 2011). In the labor-intensive small-scale farming systems, farm drudgery, higher costs of production, and rising rural wages, especially due to labor shortage caused by the male labor outmigration, are among the pressing issues that policymakers need to address for overall agricultural development, especially in the Global South. In this context, farm mechanization has been critically viewed as an important intervention to attenuate such problems of farm drudgery and labor scarcity. As such, many developing countries in Asia and Sub-Saharan Africa have prioritized farm mechanization (Aryal et al. 2019; Baudron et al. 2015; Paudel et al. 2019a). Emphasis is given to the scale-appropriate farm mechanization that is built on local agroecological and socio-economic contexts. Yet farm machines are not traditionally compatible for women due to being inappropriately designed for varied users, for example, heavyweight or improper size of machines. Moreover, the male connotation attached to machines also discourages women from taking benefit of farm mechanization (Devkota et al. 2020). In this context, attaining gender equality in farm mechanization is challenging.

In South Asia, the role and responsibility of women in household decision-making are increasing, as men, the traditional decision-makers for agriculture, are leaving this sector for taking up off-farm jobs within and outside the country (Manandhar 2021). The male labor outmigration has also increased women's burden to work in agriculture and household, on top of their already prescribed workload, fulfilling their traditional roles as home-makers and agricultural laborers. Women are involved in household activities, child and elderly care, and the added responsibility of farm management in the absence of their male counterparts, which is changing the rural labor dynamics. The intervention of farm mechanization, therefore, can be important for women to decrease farm drudgery and increase farm efficiency (Devkota et al. 2020; Paudel et al. 2019a; Pingali et al. 1987). As such, many countries in South Asia (e.g., Nepal, India, and Bangladesh) have already promulgated different farm mechanization policies, emphasizing the scale-appropriate farm machinery (Gauchan and Shrestha 2017). Yet our understanding of farm mechanization from the gender lens in South Asia is limited.

In this study, we provide novel evidence from Nepal related to farm mechanization and women's involvement in operating farm machines. Nepal is facing an acute labor shortage and rising rural wages due to rapid labor outmigration, mainly to the Gulf countries, Malaysia, South Korea, and to some extent to the western world (Khanal 2018; Wiggins and Keats 2014). A recent study shows that over 55% of the households in Nepal have at least one member who has migrated to other countries in search of better job opportunities (GoN 2020). The majority of migrating members are male household members (often the household heads), and the agricultural sector in Nepal (hills) is largely feminized (Gartaula et al. 2020). This situation has changed

Fig. 7.1 A woman operating a mini-tiller in one of the research locations. (Source: Authors)

the role and responsibility of female members to manage households and agricultural activities.

As such, in the hill ecologies of Nepal, farming is difficult due to rugged topography, fragmented land, small landholdings, and lack of access to technologies (Paudel et al. 2019b). The policymakers in Nepal are addressing these problems through the scale-appropriate farm mechanization intervention. In 2014, the Government of Nepal promulgated a farm mechanization policy (Gauchan and Shrestha 2017). The policy highlights the importance of farm machinery for rural economic development, and government programs are providing a subsidy to the farm machinery since machines are capital intensive, and the capital constraint may limit smallholder farmers (Paudel et al. 2021; Quisumbing 1995; Quisumbing et al. 2010; Quisumbing and Maluccio 2003). The government has prioritized different machines for land preparation, tillage, seeding, weeding, harvesting, threshing, and post-harvests. However, our understanding of whether these interventions are adopted by smallholders, particularly by women and marginal farmers, is limited. Therefore, in this study, we aim to understand women's responses to the intervention of mini-tillers that are widely adopted in the mid-hills of Nepal.

Mini-tillers are small 5–9 horsepower (hp) tractors (Acharya et al. 2017) that are primarily used for tillage purposes (Fig. 7.1). Mini-tillers are considered women-friendly technology due to their small size and ease of operation. This technology was introduced in Nepal in the late 2000s but was not popular due to the lack of validation in the field. After its validation in the field, especially in the mid-hills, in 2011–2013 by the research and development agencies, it has been widely spread and is seen in the hills almost everywhere nowadays. This technology is imported from

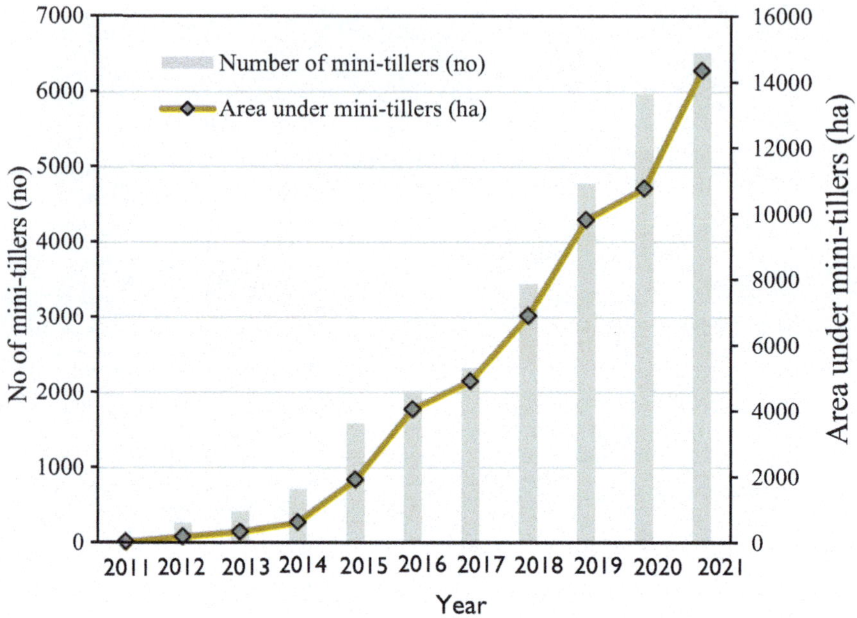

Fig. 7.2 The annual spread of mini-tillers and area under mini-tillers in the mid-hills of Nepal since its inception in 2011. Source (CSISA 2021)

China and is backed by the private sector, such as Nepal Agricultural Machinery Entrepreneur Association (NAMEA), which is involved in importing and distributing these machines through different local traders across the country. So far, over 30,000 mini-tillers have been adopted by the farmers in the mid-hills of Nepal. Figure 7.2 shows the annual trends of mini-tiller sold by the private sector NAMEA. These mini-tillers are used primarily in the maize-based farming systems in the mid-hills of Nepal. In Nepal hills, almost 0.8 million hectares of maize area are cultivated by the farmers, and labor outmigration is the common trend in Nepal hills (GoN 2020; MoALD 2020).

Policymakers in Nepal argue that this mini-tiller technology is female-friendly because of which the technology is widely adopted. There is a rich literature about the impact of this technology in the mid-hills of Nepal (Paudel et al. 2019a, b) and gender-differentiated adoption gaps on mini-tillers (Paudel et al. 2020). However, there is a lack of literature providing women's perceptions of mini-tiller technology. Therefore, here we aim to explore females' voices on this technology with the help of a farm survey conducted in the mid-hills of Nepal. The overall objective is to understand the female farmers' response to this technology and to assess the barriers that limit women's participation in the adoption of mini-tillers in the Nepal hills. The output from this research is expected to provide the gender-responsive farm mechanization programs and improve our understanding on gender and farm mechanization in Nepal.

7.2 Materials and Methods

This paper builds on the data collected from the maize-based farming system in the mid-hills of Nepal. A total of 1004 farmers were selected for the purposively sampled six districts. The details of the methodology of sample selection can also be found in some of the already published papers (Paudel et al. 2019a, b, 2020). These districts were selected based on the higher number of mini-tillers sold by the private sector Nepal Agricultural Mechanization Entrepreneurs Association (NAMEA). Among the selected samples, 179 (24%) samples were mini-tiller owners, and these owners' data is used in this analysis to understand the gender dynamics of mini-tiller adoption in the mid-hills of Nepal.

We used descriptive statistics to analyze the data used in this chapter because of the lesser participation of women in the mini-tiller adoption. Data show that only 7 (4%) sample farmers were female-headed households among the mini-tiller adopters' category. In addition to quantitative analysis of the survey data, we also analyzed data qualitatively to understand women's perception of mini-tiller operations and understand the reasons for the low level of women's participation in mini-tiller adoption.

7.3 Results

We present results for this chapter basically in four areas: (a) understanding women's participation in mini-tiller adoption, including understanding the level of and reason for (lower) adoption of mini-tillers by women farmers; (b) whether the mini-tiller technology benefits women to cater for their interests and preferences; (c) how does mini-tiller adoption reduce the farm drudgery, and (d) understanding the reasons for women's lower participation in the adoption of mini-tillers with in-depth analysis for lessons learned for policy and knowledge generation.

7.3.1 Women's Participation in Mini-Tiller Adoption

7.3.1.1 Understanding the Level of Mini-Tiller Adoption

Data show that only 4% ($n = 21$) of the female-headed households adopted mini-tillers in the maize-based farming systems in Nepal. However, when it comes to the household's decision to purchase mini-tillers, only 1% of women decided to purchase mini-tillers, while 34% of decisions were made jointly by male and female members of the households, and the rest 65% decided by men, even if they are not at home. This indicates that women farmers' decision to purchase (select) the technology is substantially lower than their male counterparts. The patriarchal culture of

Nepal has been identified as a primary reason for the low level of decision-making power for women in Nepal hills. This result has been verified by many of the previous studies (Gartaula et al. 2010; Devkota et al. 2020; Paudel et al. 2020).

7.3.1.2 Understanding Technology and Reasons for Slow Adoption

Among the mini-tiller owner households, our data shows that only one woman (0.5%) can operate the mini-tiller, indicating an insignificant level of women's participation in the mini-tiller operation. Another 2% of the owner households hire drivers for operating mini-tillers. We asked a follow-up question whether a woman can operate mini-tillers in the case where the owner hired operators. They said it needs up to three female members to operate mini-tillers, but they would not have that many neighbors in the households. This situation indicates that the female members are hesitant to use mini-tillers despite having the technical knowledge to operate mini-tillers. They know using mini-tillers is much more than just technological know-how. Altogether, only nine female members were able to handle mini-tillers.

We asked another follow-up question on what factors motivate them to learn mini-tillers operations for those women who are familiar with mini-tillers. The responses were surprising. Around 89% (n = 8) of women's motivating factors were the self-interest in mechanization technologies, while the rest started to learn the operation of mini-tillers due to lack of other family members in the households who can operate mini-tillers, showing a strong interest in the technology.

We further explored the problems faced by the female mini-tiller operators, although there were only nine female operators. Although this sample size is relatively small, the insights are important, providing broader implications for gender issues in agricultural technology adoption. The result shows that 45% (n = 4) of female operators faced problems with starting engines of the mini-tillers (Table 7.1). Even though mini-tillers are much smaller than big tractors and are considered gender-responsive technology, they still need quite a bit of force to ignite petrol or diesel and spark to run, and it requires quite some force in the beginning. This indicates that the current models of mini-tillers imported to Nepal have not considered much of the requirements and preferences of those who can exert less

Table 7.1 Difficulties faced by women mini-tiller operators in the mid-hills of Nepal

Problems	No of female operators	Percentage (%)	Cumulative percentage (%)
Taking mini-tillers to the field	$n = 3$	33	33
Engine start	$n = 4$	45	78
Difficult to control power/ speed	$n = 1$	11	89
Others[a]	$n = 1$	11	100

[a] Another problem indicates checking the frequent troubleshooting in operating mini-tillers

force on the engine. There are, however, some models where the self-starting option exists, and the introduction of such self-starting mini-tillers can solve the problem of engine start, which is an important problem women face in the mid-hills of Nepal.

Another problem we observed that women faced with mini-tiller operators were women's movement with the machine and taking mini-tillers to the field. In the Nepal hills, due to the rugged topography and lack of developed agricultural roads, farmers and operators often have to lift the entire mini-tillers to move from one place to another. Their weight ranges from 75 to 100 kg and requires multiple persons to take mini-tillers to the plots where they are used for plowing. In the lack of male laborers, it is hard for those remaining members in the household, mainly female laborers, to take machines to the field. This could perhaps be solved by developing better road connectivity, and perhaps these small plots can be consolidated to make them larger plots that facilitate the easy transportation of the mini-tillers. Other problems faced by the female mini-tiller operators were difficulties in controlling power, and frequent wear and tear of the machines (lack of local service providers for repair and maintenance), which perhaps can be solved through better women-focused training programs regarding the technical training on machines maintenance.

Many organizations are providing spare parts maintenance-related training programs specifically targeted to women. We asked a follow-up question to understand how many female operators can solve the troubleshooting problems and repair mini-tiller spare parts. The result shows that almost 55% ($n = 5$) can check the lubricant oil, repair spare parts, and fit the nuts and bolts. A follow-up question on how many female operators provided services to other farmers. They responded that no one offered such services to other farmers, indicating the absence of knowledge of business models and quality training programs for and among the female operators. These results suggest that mini-tillers are, although considered gender-responsive women-friendly machines, substantial female-targeted training programs are required to increase women's participation in mechanization and generate scaling.

7.3.2 Are the Benefits from Mini-Tiller Adoption Gendered?

We specifically investigated the differences in costs of maize production across the male-headed and female-headed households to understand whether there exists a difference in production costs of maize. This question potentially provides us with ideas on who benefits more from the technology. Results show that male- and female-headed households are equally benefited from the technology, and no significant difference for the maize inputs and outputs was detected in our data (Table 7.2). This suggests that male and female members can equally benefit from the mini-tiller adoption, although the percentage of households owning mini-tillers is substantially lower among female-headed households than male-headed households. These findings have significant policy relevance from the mechanization-based technological development perspective.

Table 7.2 Maize enterprise budgeting across male and female-headed households

Input variables	Female-headed households ($n = 7$)		Male-headed households ($n = 172$)		Significance (p-value)
	Mean	St. dev	Mean	St. dev	
Tillage costs (NPR/ha)	11,143	5113	13,675	6156	*0.28* ^{ns}
Maize seed costs (NPR/ha)	5991	3826	5755	5814	*0.91* ^{ns}
Fertilizer costs (NPR/ha)	9337	6470	6995	7672	*0.42* ^{ns}
Other cots (NPR/ha)	13,638	5129	12,245	8146	*0.65* ^{ns}
Total number of labors (no/ha)	114	32	97	37	*0.24* ^{ns}
Total variable costs (NPR/ha)	72,288	33,449	60,633	25,780	*0.24* ^{ns}
Maize yield (kg/ha)	4011	2570	3363	2228	*0.45* ^{ns}
Gross revenue (NPR/ha)	96,568	69,219	80,302	51,777	*0.42* ^{ns}
Gross margin (NPR/ha)	24,279	59,933	19,669	54,143	*0.82* ^{ns}

Note: Due to the small sample size, we analyze the ANOVA test to detect the significant difference. ^{ns} indicates not significant across male- and female-headed households

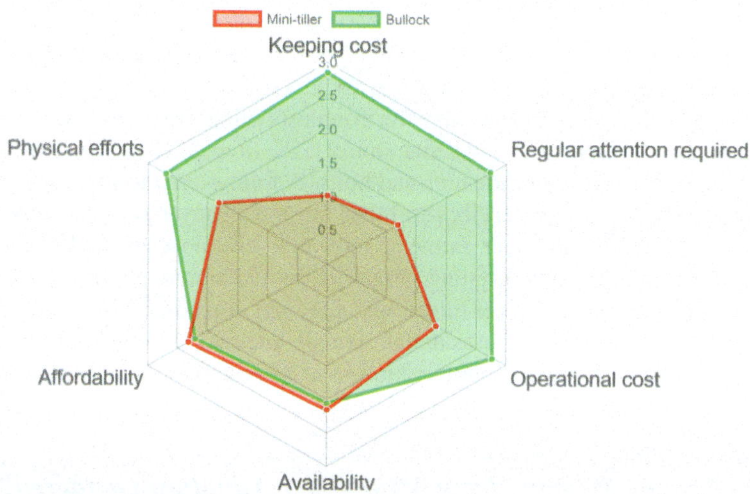

Fig. 7.3 Radar graph visualization on farmer's drudgery perception related to keeping bullock compared to mini-tiller adoption in the mid-hills of Nepal. High mean value indicate more weight [0 = low to 3 = high]

7.3.3 How Does Mini-Tiller Adoption Reduce Farm Drudgery?

We further explored how mini-tiller adoption helps farmers reduce farm drudgery in the mid-hills of Nepal. Results presented in Fig. 7.3 show that the adoption of mini-tillers substantially reduced drudgery associated with bullock keeping and traditional (or manual) farming. Traditionally bullocks and manual labors are the only sources for tillage operation and or agricultural land preparation. The cost of raising and keeping bullock is much higher due to the requirements of feeds, concentrates, and fodders and labor cost associated with raising bullock. Even though bullock provides the farmyard manure, which is useful to increase crop productivity, the higher cost of raising bullock has become a significant issue for farmers, particularly for the women farmers who must collect the fodder and grasses, often time a long walk to the forest. Women's time involvement in collecting fodder and grasses to feed their bullocks and other livestock is also a major exertion, which the mini-tiller adoption could reduce. The adoption of mini-tillers substantially reduced keeping costs for livestock, physical efforts to raise bullock, the operational cost of farming, and the attention required (mental pressure) to keep bullock. Nonetheless, affordability (high price of mini-tiller; US$ 550 per piece while compared to US$ 360 for a pair of bullocks (based on survey results)) and availability (due to remoteness) have been a major trade-off for the adoption of mini-tillers in the country.

7.4 Discussion

We observed that the benefits from scale-appropriate mechanization are not gendered while adopting technologies and embracing mechanization are. In this section, we will discuss what do we need to understand to unravel this gendered element of farm mechanization. We identified social perception, complex geography, added responsibilities for women, subsistence nature of farming, incompatible technology (inadequate understanding of technology users), and a lack of gender target programs as some of the issues explaining this complexity. We suggested some solutions to increase women's participation in agricultural mechanization.

7.4.1 Societal Perception

There is a deep-rooted societal perception in Nepal that women should not use a plow to till the soil, and if a woman is involved in soil tillage operation, the productivity of the crop declines. Although, this traditional thinking is changing over time, especially due to male labor outmigration, increased awareness, increased rural literacy, and increasing women's roles and responsibilities in farming and household decision-making. Yet, the majority of rural communities are those who still believe in socially established traditions as mottos of life. This tendency is transferred from generation to generation. It has been identified as the major barrier for women's involvement in technology adoption decisions at the household level. Generally, it also has a direct effect on how women are treated in participatory technology development. Our analysis shows that only 4% of women are involved in mini-tiller adoption decisions (Table 7.1), and we argue that these socially established traditions are responsible for this lower participation of women in the process of farm mechanization. We also find that there is no significant difference in maize productivity and farm returns across male- and female-headed households (Table 7.2). These results suggest that technology itself does not discriminate against men or women and hence there is no significant difference in maize productivity and farm returns. This also proves that given equal opportunity, women are equal to men in implementing better bet farming practices and improving farm productivity. However, how the existing social and gender norms perceive technologies is guided by the fact of who participates and who gets benefits from. Therefore, changing traditional mindset and disseminating awareness programs that tillage by male or female does not affect the crop productivity is crucial to remove this barrier, which is in addition to increasing the rural literacy in Nepal, to increase women's participation in mechanization.

7.4.2 Complex Geography

Unlikely the plain areas of the Indo-Gangetic Basin or the Nepal Terai regions, the hills of Nepal holds a rugged topography (Takeshima 2017a), small plots and fragmented land (Niroula and Thapa 2007), underdeveloped roads for transportation (Shively and Thapa 2017), and scattered settlements. Yet, over 60% of the cultivated area in Nepal is located in the hills (Ghimire and Huang 2015), primarily the maize-based farming systems. Well-developed road connectivity from farm gates to plots and at least a sufficient area of plots are required for the smooth operation of farm equipment. Nepal hills lack such conditions, limiting women's involvement in farm machinery adoption. Moreover, many farm machines are heavy, and in the absence of better road connectivity and smaller land parcels, these machines sometimes need physical lifting to move them from one plot to another. These machines are too heavy for a single person. They need support from other household or non-household members, which ultimately limits the adoption of farm machinery in general but for women, in particular. We suggest that the government programs in Nepal should focus on developing rural roads for better transportation. Consolidating several plots into one could be another option, but in the hills, it would pose another challenge by creating a slope issue. Nevertheless, focusing on the design of smaller mini-tillers and developing smaller tracts for better movement of those smaller machines would be another alternative to increase the efficiency of machines and get better women's involvement.

7.4.3 Added Responsibilities

In Nepal, because of male labor outmigration, the management of rural households and farmlands has been in the hands of women (Doss et al. 2022; Lokshin and Glinskaya 2009; Tuladhar et al. 2014). The women, children, and elderly have been shouldering on agricultural performance, and households are becoming female-headed, even if they are de-facto household heads. In the absence of male members in the households, women have to handle responsibilities, including cooking food, child and elderly care, and off- and on-farm economic activities. The activities that were handled by men are now additional work for women, increasing their labor burden. This added responsibility limits their participation in mechanization because the machine needs specific skills and they do not have time to learn these new skills. Due to higher workloads, they are not able to focus on machines and technicalities associated with their operation. The government is promoting the custom hiring services in Nepal (Takeshima 2017b), where farmers can rent the services on an affordable pay-for-services. The women can make use of this opportunity saving their time and reducing the overloads from the additional responsibilities which need to be taken care of due to male migration.

7.4.4 Subsistence Farming

Agriculture in Nepal is characterized by subsistence farming in a mixed farming system where smallholder farmers grow crops, vegetables, fruits and keep some small and large livestock generally for home consumption. Cultivating for market and commercial purposes is very limited. While the agriculture in the lowland areas (e.g., Nepal Terai) is leading towards commercialization due to better road connectivity, market access, high access to irrigation (e.g., groundwater), plain areas, and large plots size, hilly areas has limited access to agri-inputs, markets, and small plots with rugged topography. Farmers, therefore, in hilly areas grow crops for their own household consumption, and they are not oriented towards the market. When farmers are oriented towards the market, they are likely to use machines that increase farm efficiency and productivity. Significant development programs are required to change the mindsets of female and male farmers to change their subsistence to commercial farming. The Government of Nepal has launched several programs such as Agricultural Perspective Plan (APP) Agriculture Development Strategy (ADS) to change the existing subsistence farming to commercial; many programs failed to fulfill the objective. The smallholder farmers do not have much enthusiasm and incentive to go commercial due to the factors mentioned above, which also contribute to women's less participation in improved technology programs and machine trainings. Hence, farmers have to be aware of the opportunities lying around for their growth and make proper use of the opportunities offered by the government or non-government agencies working for farmers.

7.4.5 Technological Incompatibility

At present, Nepal has no manufacturing plants to produce farm machinery, and it imports all of them from neighboring countries, mostly from India and China. The imported machines are not specifically designed by looking at the Nepalese market, let alone the tailored design of looking at the requirements and preferences of diverse end-users, including women. The private sector imports these technologies by self-evaluating and looking at the local agroecological contexts. Hence, most of the imported machines do not fit the local requirements. Our results show that many of the mini-tillers are difficult to operate, especially for women who face problems to kickstart the machines, heavyweights, and improper size. This has a great potential to limit women's participation in farm machinery.

One way to address this issue is that the government establish farm machinery training and testing centers in the country so that the imported machines are first tested and validated in the research stations before they are recommended for adoption. Moreover, if the machines are not suitable, they should not recommend such machines for wider scaling. Currently, due to a lack of sufficient training centers, the government is unable to conduct a test of such machines. The government could establish such training and testing facilities through Nepal Agriculture Research Council (NARC) – an apex body to conduct research in Nepal. This could help identify the appropriate farm machines that are suitable for women in the hills and could increase women's machinery adoption.

7.4.6 Gender-Responsive Capacity-Building Programming

The results show that women face problems related to operating a machine, such as fitting the spare parts, power control problems, and difficulty in engine start. These problems can be addressed by providing the skills training targeted to women as new or emerging clientele groups. The training program can be specifically conducted by the government or non-governmental development agencies. Although the government is focusing on a wider spread of farm machinery to handle the problem of labor scarcity, there is a limited focus as far as skill development of farmers related to spare parts maintenance, machine operation, and repairs training are concerned. We argue that the government and agriculture development agencies in Nepal should specifically target female members in the rural areas to provide them with need-based capacity development and training programs which will increase the machinery knowledge to both female and male farmers. Hence, it is time to focus on increasing the programs related to the capacity-building of women farmers.

7.5 Conclusion

The global sustainable development goals (SDGs) can only be achieved by providing required attention to the gender-responsive scale-appropriate mechanisms, especially in the agricultural sector. In this paper, we analyzed a subset of data from an extensive survey on mini-tiller adoption in the hills of Nepal whether the scale-appropriate farm mechanization is gender-sensitive. Apparently, the scale-appropriate farm mechanization has been getting momentum in recent years as a solution to the problem of labor scarcity in the rural areas, which is largely created due to male labor outmigration. However, the understanding is limited when it comes to gender dimensions of farm mechanization. We observed that only 4% of the owners were female farmers, and only 1% of them were involved in mini-tiller operations.

Moreover, we found that maize productivity and farm returns did not vary across male and female-operated plots. The chapter identifies the social perception against women for tillage operation, women's low level of knowledge in operating machines, spare parts availability and maintenance, added responsibilities, and workload due to men's migration of the household as the major factors responsible for women's lower participation in mini-tiller adoption. The paper concludes with some recommendations to increase the adoption of mini-tillers by women. There are increasing needs for awareness-raising, increasing access, and designing training programs focused on women and built around tailoring women's requirements and subjective social conditions. This will eventually have a bigger impact on maize productivity more equitably and contribute to SDG by women's empowerment through technical skills.

Acknowledgments This study was conducted as part of the project "Cereal System Initiatives for South Asia" (CSISA). The project is supported by the United States Agency for International Development (USAID) and the Bill and Melinda Gates Foundation (BMGF). We gratefully thank the farmers for their time in participating in the survey and Anish Dahal, Bivek Aacharya, Kiran Dahal, and Sameer Shrestha for conducting the interviews with farmers. This paper reflects the authors' views and not those of funders and authors' institutions.

References

Acharya R, Budhatohki-Thapa PK, Khanal P, Maharjan S, Syangtang S (2017) Technology profile: mini-tiller. Integrating gender and nutrition within agricultural extension services (INGENAES) and USAID feed the future, Nepal

Aryal JP, Rahut DB, Maharjan S, Erenstein O (2019) Understanding factors associated with agricultural mechanization: a Bangladesh case. World Dev Perspect 13:1–9

Baudron F, Sims B, Justice S, Kahan DG, Rose R, Mkomwa S, Kaumbutho P, Sariah J, Nazare R, Moges G, Gérard B (2015) Re-examining appropriate mechanization in eastern and southern Africa: two-wheel tractors, conservation agriculture, and private sector involvement. Food Secur 7:889–904

CSISA (2021) Cereal systems initiative for South Asia. Phase III. Annual report. International Maize and Wheat Improvement Center (CIMMYT), Kathmandu, Nepal

Devkota R, Pant LP, Gartaula HN, Patel K, Gauchan D, Hambly-Odame H, Thapa B, Raizada MN (2020) Responsible agricultural mechanization innovation for the sustainable development of Nepal's hillside farming system. Sustainability 12. https://doi.org/10.3390/SU12010374

Doss CR, Meinzen-Dick R, Pereira A, Pradhan R (2022) Women's empowerment, extended families and male migration in Nepal: insights from mixed methods analysis. J Rural Stud 90: 13–25. https://doi.org/10.1016/j.jrurstud.2022.01.003

FAO (2011) The role of women in agriculture, ESA Working Paper No. 11-02, Agricultural Development Economics Division

Gartaula HN, Niehof A, Visser L (2010) Feminization of agriculture as an effect of male outmigration: unexpected outcomes from Jhapa District, Eastern Nepal. Int J Interdiscip Soc Sci 5(2):565–577

Gartaula H, Sapkota T, Khatri-Chhetri A, Prasad G, Badstue L (2020) Gendered impacts of greenhouse gas mitigation options for rice cultivation in India. Clim Chang 163:1045–1063

Gauchan D, Shrestha S (2017) Agricultural and rural mechanisation in Nepal: status, issues and options for future. In: Mandal SMA, Biggs SD, Justice SE (eds) Rural mechanisation. A driver in agricultural change and rural development. Institute for Inclusive Finance and Development, Dhaka, pp 97–118

Ghimire R, Huang W (2015) Household wealth and adoption of improved maize varieties in Nepal: a double-hurdle approach. Food Secur 7:1321–1335

GoN (2020) Nepal labour migration report 2020. Ministry of Labor, Employment and Social Security, Government of Nepal

Khanal U (2018) Why are farmers keeping cultivatable lands fallow even though there is food scarcity in Nepal? Food Secur 10:603–614

Lokshin M, Glinskaya E (2009) The effect of male migration on employment patterns of women in Nepal. World Bank Econ Rev 23:481–507. https://doi.org/10.1093/wber/lhp011

Manandhar S (2021) Women representation in Nepalese local election 2017: issues and challenges. Aust J Polit Sci 21:34–48. https://doi.org/10.3126/jps.v21i1.39283

MoALD (2020) Statistical information on Nepalese agriculture. Ministry of Agriculture and Livestock Development, Government of Nepal. Singhdurbar, Kathmandu, Nepal

Niroula GS, Thapa GB (2007) Impacts of land fragmentation on input use, crop yield and production efficiency in the mountains of Nepal. L Degrad Dev 18:237–248. https://doi.org/10.1002/ldr

Paudel GP, Gartaula H, Bahadur D, Craufurd P (2020) Gender differentiated small-scale farm mechanization in Nepal hills: an application of exogenous switching treatment regression. Technol Soc 61:101250

Paudel GP, KC DB, Rahut DB, Justice SE, McDonald AJ (2019a) Scale-appropriate mechanization impacts on productivity among smallholders: evidence from rice systems in the mid-hills of Nepal. Land Use Policy 85:104–113

Paudel GP, KC DB, Rahut DB, Khanal NP, Justice SE, McDonald AJ (2019b) Smallholder farmers' willingness to pay for scale-appropriate farm mechanization: evidence from the mid-hills of Nepal. Technol Soc 59. https://doi.org/10.1016/j.techsoc.2019.101196

Paudel GP, Timsina KP, Karki S, Adhikari S, Babu TSA, Krupnik TJ (2021) Slow adoption of mechanical rice transplanters in Nepal's cereal-based cropping systems: reasons and recommendations for change. The Cereal Systems Initiative for South Asia (CSISA), Kathmandu

Pingali P, Bigot Y, Binswanger HP (1987) Agricultural mechanization and the evolution of farming systems in Sub-Saharan Africa, World Bank. The Johns Hopkins University Press Baltimore, Maryland, USA

Quisumbing AR (1995) Gender differences in agricultural productivity. FCND Discussion Paper. International Food Policy Research Institute, Washington

Quisumbing AR, Maluccio JA (2003) Resources at marriage and intrahousehold allocation: evidence from Bangladesh, Ethiopia, Indonesia, and South Africa. Oxf Bull Econ Stat 65:283–327

Quisumbing AR, Pandolfelli L, Brook S, Brook S (2010) Promising approaches to address the needs of poor female farmers: resources, constraints, and interventions. World Dev 38:581–592
Shively G, Thapa G (2017) Markets, transportation infrastructure, and food prices in Nepal. Am J Agric Econ 99:660–682. https://doi.org/10.1093/ajae/aaw086
Takeshima H (2017a) Overview of the evolution of agricultural mechanization in Nepal. IFPRI Discussion Paper 01662, Development Strategy and Governance Division. International Food Policy Research Institute, Washington, DC
Takeshima H (2017b) Custom-hired tractor services and returns to scale in smallholder agriculture: a production function approach. Agric Econ 48:363–372
Tuladhar R, Sapkota C, Adhikari N (2014) Effects of migration and remittance income on Nepal's agriculture yield. No. 24. ADB South Asia Working Paper Series, Asian Development Bank, Manila, Philippines
UN (2015) Transforming our world: the 2030 agenda for sustainable development, United Nations General Assembly
Wiggins S, Keats S (2014) Rural wages in Asia. Overseas Development Institute 203 Blackfriars Road London SE1 8NJ. UK

Chapter 8
Land Rights of Women in India: How Much Has Changed After 2005

Anuradha Rai

Abstract Gender equality has been the cornerstone of the Indian constitution and India's democratic principles; however, gender inequality and biases persist in Indian society and the legal system of the country at large. In the arena of inheritance and succession especially, women have been deprived from making any claim on land and property. As most of the land in India is passed through inheritance, very few women in India own land in a patriarchal society where gender biases are deeply enrooted in the very social fabric and culture of the society. There have been many studies highlighting the inter-relationship between women's right to land and property and women's security saying that it leads to constitute a complex situation for women and exposes her for physical and structural violence. Keeping the same in the preview, Indian government has revised the land and inheritance law from time to time to ensure women's right over land and other resources. In a major breakthrough, in the year 2005, the government has amended the existing Hindu Succession Act 1956 to guarantee women equal right with men in land and ancestral property. However, little has changed on ground. The proposed paper will analyze the Hindu Succession (Amendment) Act 2005 and its effectiveness for women's claim to land and property in India.

Keywords Women · Inheritance · Land · Property · Violence

8.1 Introduction

Throughout the world, historically, women had been considered as property themselves who should be controlled, monitored, and mended by the father, brothers, and husband, in general terms by the men of the society. Hence, a fight to own and inherit land and property for women had been a long and difficult battle that went much longer than any great battles in the world. The battle is still ongoing in many of the

---

A. Rai (✉)
Department of Political Science, Assistant Professor, Jananayak Chandrashekhar University, Ballia, India

S. Munshi, M. Singh (eds.), *Women Farmers: Unheard Being Heard*, Sustainable Agriculture and Food Security, https://doi.org/10.1007/978-981-19-6978-2_8

Asian, African countries and also rest of the world with India being no exception. Though, unlike other parts of the world, Indian women have the privilege to constitutionally get their political and economic rights, the ownership of land to women is something the Indian society barely acknowledge. Throughout the world, gender inequality, when it comes to land and other productive resources, is intimately related to women's poverty and exclusion from ownership of one of the basic resources of living that is land.

According to Bina Agarwal (2005, 2021), the social and economic implications of denial of women right to inheritance and ownership of land are wide-ranging. The livelihood of millions of women in Asia, Africa, and Latin America is critically dependent on land. According to her, the typical process of agrarian transformation around the globe under which labour shifts from agriculture to non-agriculture has been slow and gender-biased. In many countries, those who have moved from agriculture to non-farm work are largely men, whereas women have remained substantially in agriculture.

Land is not only an important asset but a critical resource, especially for women, as women comprise around 43 percent of total agricultural labour force globally and just over 40 percent of total agricultural labour force in developing world (FAO 2011). As per the statistics, in India, for instance, 58 percent of all male workers, 78 percent of all female workers, and 86 percent of all rural female workers, are in agriculture. Even though women comprise around half of the total agricultural labour force, less than 20 percent of landholders around the world are women (FAO 2010). In South Asia, 35 percent of total agricultural labour force and more than 70 percent of female labour force is engaged in agriculture but only around 10 percent own total landholdings (FAO Fact Sheet 2010–2011). The condition of women in South Asia is even worse as women in most of the South Asian countries have legal rights to own and inherit land, but the legal contradictions and the biased customary practices dominate and prohibit women's right to demand their own land. Indian women are no exception to these treatments where women at large face discrimination in social, political, and economic arenas, land being one of it where women are traditionally denied any ownership of land.

Indeed the lack of ownership of land to women has been many a times triggering factor behind the struggle faced by women who suffers rejection, divorce, separation, and violence from the family, husband, and men of the family, In late 2005 the government of India took a cognizance and there was a legal and political response to address the issue taking initiative to change the Land Law of 1956 (Gazette of India).

The land law of 2005 is considered as a major leap towards granting equal inheritance rights to women in accordance to the fundamental rights of equality guaranteed by the Indian constitution. This brought for Hindu women, who constitute 80% of women in India, substantial legal equality in all forms of property, including agricultural land, and gave all daughters (married and unmarried) coparcenary rights by birth in paternal joint family property (Agarwal 2005). It was presumed that once the legal rights to inheritance and property for women would be established, it will bring major change at the societal level also.

However, no drastic change has been registered in the last 15 years after the implementation of the new Land Law of 2005. This also has a direct consequence as far as women in agriculture is concerned. In India, almost one-third of cultivators are women, but they own less than 10.34 percent of land, operating 12.8 percent of holdings, as per Agriculture Census, 2010–2011, while 75 percent of female workforce, largely marginal or landless, depends on agriculture for survival. The average size of women's land holding in India is 0.93 ha, in comparison to 1.18 ha for male and 1.15 ha for all (Chowdhry 2017).

8.2 What Is "New" in the New Land Law of 2005

The framers of the Indian Constitution took note of the adverse condition of women in Indian society and a number of provisions and safeguards were included in the Constitution to address the problem of gender inequality. In this context, Articles 14, 15(3) and 16 of the Constitution can be mentioned. Article 14 of the Constitution of India states that: "The State shall not deny to any person equality before the law or the equal protection of the laws within the territory of India". "Article 15: Prohibition of discrimination on the grounds of religion, race, caste, sex, place of birth or any of them. These provisions are part of the Fundamental Rights guaranteed by the Constitution. Part IV containing Directive Principles of State Policy, which are no less fundamental in the governance of the State to ensure equality between man and woman such as equal pay for equal work. The Directive Principles further endorses the principle of gender equality, which the State has to follow in matters of governance.

In India, beyond the constitutional guarantee of gender equality, the Hindu Succession Act of 1956 (HSA 1956) is the fundamental law governing present day inheritance rights of four religions: namely Hindus, Buddhists, Jains, and Sikhs. According to the Act of 1956, when a male Hindu dies after the Act came into force, his interest in a Mitakshara ("inheritance by birth") coparcenary shall go to the surviving members of the coparcenary and not in accordance with the Act. However, a provision was added to preserve the interest of female children. It said if the deceased left behind a Class I female relative (daughter, widow, mother etc.) or a male relative claiming through such female kin, his interest would go to them by testamentary (by will) or intestate (without a will) succession, and not by survivorship. According to this law, daughters did not have equal inheritance rights to the family property. As per this law, daughters could jointly inherit property acquired directly by their fathers, but ancestral property could only be inherited by sons. The prevailing biased clauses hence created an urgent need for the amendment of the act making it more gender sensitive.

More than 90 percent of agricultural land in India is transferred through inheritance which is passed from the father to sons. Henceforth, Land and especially the agricultural land remained out of women's right to inherit and own. To address the problem of inheritance discrimination against women, the HSA 1956 law was amended between the 1970s and 1990s by five Indian states of Andhra Pradesh, Tamil Nadu, Kerala, Karnataka, and Maharashtra to allow the unmarried women, at the time the reform was passed, to inherit ancestral property.

The 2005 Hindu Succession Act was legislated targeting the goal of women empowerment by giving them equal share in the ancestral property to women including land, a major resource on which women in India had no authority. While noting down the discrimination against women in land and property inheritance right, and proposing the changes made in the Act of 2005, the Union government had argued that, "The law by excluding the daughter from participating in the coparcenary ownership not only contributes to her discrimination on the ground of gender but also has led to oppression and negation of her fundamental right of equality guaranteed by the constitution. . . . . ." (Hindu Succession Act 2005) The control over land in India is primarily in the hands of men though, women comprise a major workforce in agricultural sector. According to OXFAM India report, agriculture sector employs 80 percent of all economically active women in India. Women comprises 33 percent of the agricultural labour force and 48 percent of self-employed farmers. In rural India, 85 percent of women are engaged in agriculture, yet only 13 percent of women own land.

The Hindu Succession (Amendment) Act (HSAA) 2005 provides for four clear rights to women:

1. Women are coparceners at par with men.
2. A woman can inherit agricultural land from her parents at par with her brothers.
3. A woman can also inherit the self-acquired agricultural land of her deceased husband; and
4. A widow who is remarrying can inherit the land of her deceased erstwhile husband.

The Act of 2005 has made significant change in the Hindu succession Act by making daughters (including the married daughters) coparceners in joint family property. It has both economical and symbolic importance. Economically, it has been presumed that by giving birth right to women in the land and property, the security to women will enhance. In case of failure of marriage or infliction of violence women can return to her birth place and can live on her own; not on the mercy of her relatives. Symbolically, in a male-biased society, where birth of a girl child is looked down the Act signals that both daughters and sons are equally worthy and important members of the family. It undermines the notion of daughters belonging to some other family, paraya dhan (husband's family asset) and outsider to the parental family.

8.3 The New Land Law and Women in India: An Assessment

The New Land Law has brought a powerful ray of hope as far as discrimination to women in context of land rights are concerned. Where on the basis of their gender they were discriminated earlier the laws act as support for millions of women to claim their due rights and have a life of dignity. This has also created a conducive platform for women to avail their land rights. Along with this positive development, there are still various struggles, formal and informal barriers as far as the practical implementation of the clauses are concerned. Understanding those barriers will help in increasing the efficient implemenations.

Mira, a resident of Uttar Pradesh, is 43 years old women married at the age of 19 and has one daughter. Her husband lives and works in Delhi. Mira was living with her in-laws taking care of the family and the child. Eight years back, after 16 years of marriage, her husband announced his marriage with another women in Delhi and asked Mira to leave his house. Mira's brothers after knowing the circumstances, brought her to her parental home. Mira has no formal education and can hardly understand the legal course of actions required to take any action against the injustice done to her by her husband. She has no job to earn her living and is forced to live with her mother and brothers. Her brothers are busy in their own lives and work and they can hardly bear the expense and sufferings of legal battle for Mira.

As Mira has only one daughter from her marriage, they think that they will get the girl married and she will leave for her husband's home. For Mira, she has no other option but to live on the mercy of her brothers doing the domestic chores and working on the farms of the family. Though, Mira is working on the lands of her family, she has no claim on the land and earnings. She has no place of her own and no land of her own. She apprehends about her life after her mother's death and fearful of how her brothers and sisters-in-law will behave in her old age. Talking about her fears for the future she tells, "I have no place of my own, no home and no certainty of my future. What if my brothers one day throw me out of their house after my mother? Only if I would have had a son, my situations might would have been better as he would have had claim in his father's home and farms" (Interviewed on 17 November 2020). In the case of Mira, what would have been the response of her brothers if she would have had a son and not daughter? In the absence of any source of earning and money, can Mira fight her case in her in-law's house. Can Mira take the risk of claiming her right to inherit land in the parental property and lose the social sympathy and support of her family to upbring her daughter? Can she even think of going to the court of law to put her claim? These are the questions she asked when she was briefed about the New Land Law of 2005.

Pinki is far stronger than Mira and knows that the legal position of the girls has changed regarding the land ownership. Therefore, when her father died in 2014 and her brothers and mother had clashes regarding family property paying no attention to Pinki's health and education, she filed her case for separation and got her share of land on her name which led to severe criticism from family, relatives, and society leading further to social boycott of Pinki. Since then, Pinki is all on her own and is isolated and she feels completely alone and insecure. Narrating her sufferings, she opened up, "nobody came to ask me how I am surviving when my sister-in-law denied any food and my daily requirements. My mother went to her parental home for six months and I was all alone thinking about my life after my father died. When I told my mother about my problems, she ignorantly said that she has her own challenges of life that she does not know how to handle. Finally, one day, I decided to claim my right to land to be able to take care of myself and not look around for help. Since then, I became an enemy and a threat to everyone. People started coming to convince me that I have done wrong with my family" (Interviewed on 21 April 2021).

Pinki has formal education till secondary and can read and write but with no permanent source of income to make her living. She is forced to sell her land for living and she signs the papers after thoroughly going through it. Though, she says that she can understand little the legal terms and conditions. She is also facing difficulty in getting the money from the buyers as the middlemen takes a bigger share of her money. She is totally broken and says that she does not know if she has made a mistake by taking her share and claiming her right to property. She says that she gets no proposals of marriage and she has become infamous among the people around her. One man, a lawyer, has recently proposed her for marriage but she is not sure about his intentions. She is fearful that the man might be interested in her property and not her as his family does not like her at all. She feels insecure about her life and says no one will pity her if she is killed or she dies. She feels blank and finds no future for her in a completely hostile society.

These cases reflect upon the dire need for equal social responsibility which is needed to give practical shape for achieving equity and equality. These are among hundreds of the cases that women in India are still facing despite the changes made in the legal system of inheritance and ownership of property in their favor. The above cases also indicated the sufferings and hindrances to women in making their claim to land in Indian society. According to the agricultural census 2001, 10.8 percent of women used to own land in India. The 2010–2011 census of the Government of India show a marginal increase with 10.34 percent of women owning land in India which as per the Agricultural census of 2015–2016 increased to 11.57 percent of total landholdings. These statistics indicates a marginal change in the condition of women in case of ownership of land and agricultural landholdings in India.

In December 2013, International Land Coalition (ILC) member Landesa Rural Development Institute and UN Women, India, prepared a report on the formal and informal barriers in the implementation of the Hindu Succession (Amendment) Act, 2005 in the context of women agricultural producers of Andhra Pradesh, Bihar, and Madhya Pradesh. The studies conducted reveal several barriers to the implementation of the Land law of 2005. The report notes that:

1. Hindu social practice allows the parents to free themselves from honouring the daughter's inheritance right at par with their sons with the excuse of paying dowry at the time of the daughter's marriage. Dowry as a substitute of land and

other properties in inheritance is one keyway the patriarchal beliefs are deeply anchored in social practice, denying the women social and economic equality within the family. The practice of dowry is a penal offence but is so entrenched that women themselves do not feel that it is their moral or legal right to claim inheritance rights in their parents' property.

2. Women's understandings of the current inheritance law and of the claim processes are clearly rudimentary and fragmented. Land has historically been a male domain, and it continues to be so.
3. Even when the women do receive land in inheritance, it is invariably much less than an equal share. Women are likely to get more land as widows than as daughters.
4. The people and institutions that are mandated to enforce the law are prisoners of the same practice.
5. Even when women do get land in their own names by inheritance, through dowry, or through purchase by their marital family in their name (this is often done to take advantage of reduced stamp duty on property purchased in a woman's name), the ownership by women is only notional. The women are seldom in possession of the land, title, and the Record of Rights (ROR) that make it a secured tenure. The decision-making power on use of the land remains firmly in the grip of men of the family—father, brother, son, husband, or father-in-law.
6. There is a lack of political will on the part of the State governments to implement the law.

Similar study has been done by Nitya Rao (2018) on women and land ownership in Jharkhand region indicating hurdles women of the particular state and India in general are facing to claim their right to land as per the Land Law of 2005. According to her, the social perception of women who claim their right to land is always negative and discouraging. According to her, "women have been an integral part of the movement for political autonomy in Jharkhand, since the 1960s but the issue of their independent land rights has been actively opposed, more so, in the last decade. Women publicly asserting their rights to land have been branded as "dains" (witch), harassed, and killed., Land ownership right which is about recognition and identity is often interpreted as a greed and more so a challenge to male authority. It is seen to be disruptive of the social order and hence needs to be controlled". More so it is understood to be a challenge to the long-time patriarchal power game. It not only challenges the stereotypical social perception of women being a dependent and men the saviour rather it also challenges the so-called manhood which is entrenched in the deep-rooted patriarchal social structure.

According to these and other studies conducted in different regions of the country, although women now enjoy legal support to claim their equal right, there are various restrictions noted down that hampers the women's right to land. The very first among those is the gender-biased social ecosystem which questions women's claim to land as a normal affair. The customary practices, social pressure dominate the legal provisions. People relates customary practices with their cultural identity and bypass the laws to protect and preserve it. Women often considered as the sole caretaker of the traditions and customs many a time suffers on the hand of the existing social norms which restricts her to seek justice, claim her rights on the name of traditions.

In most part of the country, it is considered to be shameful for women to claim her right in the court with her male family members and women has to face social boycott and rejection as happened in the case of Pinki.

Another important restriction on women right to land is that women often renounce their statutory right to land due to her economic and social dependence on her kins. In most part of the country, there is strict demarcation between private and public arena, restricting women to enter the public arena. The socialization of women since childhood is done based on the existing discriminatory practices. This results women's dependency on the male members of the family, leading her to be cautious about not challenging the customary practices. A study by UN Women and Landesa found that 44 per cent of the women surveyed believed their parents would not agree to give them a share of their land. While 53 per cent believed their brothers would oppose the idea.

Indian women are hesitant to claim their right to inheritance due to lack of education and awareness. With primary education or no education at all, she lacks the confidence to claim her right and understand the legal complications of her claim. Ten years after the Right to Education (RTE) Act came into being, nearly 40% of adolescent girls in the age group of 15–18 years are not attending school while 30% of girls from poorest families have never set foot in a classroom, according to the report by Right to Education Forum and Centre for Budget Policy Studies with support of World Bank and UNICEF (TOI Report, 25 January 2020). Women being uneducated puts her at the risk of poverty and violence. With few women attending the higher education, the chances of women getting the self-confidence and courage to reach the court of justice and understand the legal position of their claim to land is thin.

The lack of mobility of women in India and her absence from public arena is another important factor limiting her horizon to access the land rights legal support. With limited exposure owing the restrictive movements she is unable to have access to the legal aspect of her right to land. Furthermore, as women's land right in India is not uniform and is governed by various personal laws and customary practices, there is lot of misunderstanding about the exact position of women's right of inheritance. Shruti Pandey, a PIL lawyer in the Supreme court argues that the property right of Indian women depends on which religion she follows, her marital status, the state she belongs to and whether she is a tribal or non-tribal etc. which creates confusion and complication for most of the women in India to understand her legal position to inheritance making them hesitant to claim her right to land (Mehta 2018).

8.4 Conclusion

The Economic Survey 2017–2018 says that with growing rural to urban migration by men, there is "feminization" of agriculture sector, with increasing number of women in multiple roles as cultivators, entrepreneurs, and labourers (OXFAM 2018). However, few women have their claim over the land they work onto. Though, the reforms in the inheritance and property law to bring gender justice in India started soon after independence, with passing of Hindu Succession Act 1956, women in India are still at margins when it comes to land right of women. The marginal increase in the landholdings of women cannot be considered as a result of women's increasing share in the parental property but a mix of women's increase in the share of marital family, lower stamp duty on registering land on the name of women and other such reasons. The New Land Law is a positive development in favour of women and there are needs for its quality implementation.

With the increasing feminization of agriculture in India, it is imperative to implement mass land rights awareness campaign in rural India. The various clauses need to be explained to the women where she has access to the legal advocacy at her doorsteps, if she wants to claim the right. At the community level too, it is imperative to generate awareness and shun off the myth related to the New Land Law. Women who have taken stands about accessing the New Land Law should be awarded or recognized at the government level so that others also have courage to take a stand. The land rights of women is yet another arena which needs constructive interference both at social and political level to strengthen the constitutional values of equality and justice. Land rights of women will act as a catalyst in the next green revolution journey, where her land ownership will also create a level playing field as far as decision-making is concerned. Last but not least it is a shared responsibility of the social ecosystem for making the New Land Law a success at the grassroots.

References

Agarwal B (2005, September 25) Landmark step to gender equality, The Hindu. http://www.binaagarwal.com/popular%20writings/Landmark%20Step%20to%20Gender%20Equality%20TheHindu_25sep05.pdf

Agarwal B (2021) How many and which women own land in India? Inter-gender and intra-gender gaps. J Dev Stud. https://doi.org/10.1080/00220388.2021.1887478. Accessed 9 Aug 2021

Chowdhry P (ed) (2017) Women's land rights: Gender discrimination in ownership (Land Reforms in India series), Sage Publications, New Delhi

Rao N (2018) Good women do not inherit land: politics of land and gender in India. Routledge, London

FAO (2011) Women in Agriculture: closing the gender gap for development, The State of Food and Agriculture, Rome

Government of India (1956) The Hindu Succession Act 1956. The Gazette of India

Natural Justice, Indian Women's Rights to Property: Implementation of the Hindu Succession (Amendment) Act (2005)
Mehta A (2018) Gender gap in land ownership. https://www.ncaer.org/news/gender-gap-in-land-ownership#flipbook-df_manual_book/1/. Accessed 22 Mar 2021
The Gazette of India (n.d.) The Hindu Succession (Amendment) Act 2005. The Gazette of India
OXFAM India (2018) Move over 'Sons of the soil': why you need to know the female farmers that are revolutionizing agriculture in India. URL: https://www.oxfamindia.org/women-empowerment-india-farmers. Accessed 8 Aug 2021

Chapter 9
Transformative Approaches to Empower Tribal Farmwomen of Ladakh Cold Arid Himalayan Deserts: Challenges and Solutions

Shahnawaz Rasool Dar, Nazir Ahmad Ganai, Shabber Hussain, Mohammad Yaqoob Wani, Aabid Hussain Lone, Showket Ahmad Dar, Syed Berjes Zehra, Tashi Dolkar, Tashi Dawa, Rajbir Singh, Sameera Qayoom, P. S. Pandey, Muzaffar Hussain Dar, Alamgir Ahmad Dar, Ishrat Jan, Javeed Ahmad Sofi, Irshad Hassan Dar, Rehana Rasool Bhat, and Khursheed Ahmad Dar

Abstract Agriculture on the mountains is going through unprecedented difficulties like resource exhaustion, adverse environmental degradation, and shifting of youth to cities in search of jobs. Collectively, all these challenges pose thoughtful threats to food security, sustainability, and livelihoods of tribal farming communities of Himalaya and women farmers of region in particular. Hunger and chronic

S. R. Dar (✉)
RCRQA, Dean Faculty of Horticulture, SKUAST-Kashmir, India
e-mail: srdar@skuastkashmir.ac.in

N. A. Ganai · A. A. Dar · I. Jan · J. A. Sofi · I. H. Dar
SKUAST-Kashmir, India

S. Hussain · M. Y. Wani · A. H. Lone · S. A. Dar · S. B. Zehra · T. Dolkar · T. Dawa
KVK Kargil II Zanskar, SKUAST-Kashmir, India

R. Singh
Zonal Project Director, Zone I ATARI, Ludhiana, India

S. Qayoom
Head Agromet Unit, SKUAST-Kashmir, India

P. S. Pandey
ADG, EP & HS, ICAR, New Delhi, India

M. H. Dar
Department of Basic Science & Humanities, SKUAST-Kashmir, India

R. R. Bhat
Division of Soil Science, Faculty of Horticulture Shalimar, SKUAST-Kashmir, India

K. A. Dar
Division of Agronomy, Faculty of Agriculture Wadura, SKUAST-Kashmir, India

S. Munshi, M. Singh (eds.), *Women Farmers: Unheard Being Heard*, Sustainable Agriculture and Food Security, https://doi.org/10.1007/978-981-19-6978-2_9

undernourishment are on the rise; concurrent with agriculture productivity gains across other parts of nation, tribal farming community of mountainous region are worst sufferers. Rural tribal women must play very versatile role in all the major agricultural activities to ensure food security and improving the economic status of the farm family. Even though women contributes massive labor involvement in this sector, women's contribution frequently stays unseen and unappreciate. There is a need for creating equal opportunities for women across agriculture food system. In perspective of feminization of agriculture lack of low female literacy has grim implications on future productivity of agriculture sector and food security. Poorly educated rural women are likely to meet the worst effects of structural changes in economy, mainly in agriculture sector which is getting oriented towards competitive international market in the current time. Very little is known and documented about social dimensions of technology access and acceptance. Scientific advances, technological interventions, and use of digital technologies have vast potential to contribute to a larger scale structural transformation of food systems in cold arid deserts of Ladakh. Research programs needs to prioritize the understanding of socio-economic dimensions related to techonolgies being adopted by the women in agriculture and perspective who Government led institutes like KVKs can be path finder in recognizing the efforts and needs of women farmers.

Keywords Mountains · Food security · Farmwoman · Education · Technology · Access

9.1 Introduction

Faster economic progress during the last decade happened in most developing countries, the subsequent gains in income have been frequently unevenly distributed. As a result, this growth has never ensued in proportionate poverty decrease for the most susceptible groups (World Economic and Social Survey 2011). Despite essential measures grounded on scientific development, like introduction of better yielding varieties and integrated farming approaches, malnourishment is still a challenge that the globe, exclusively emerging countries like India cannot overlook.

Mountain agriculture is physically challenging and time consuming. Rural women have vital roles as primary food growers and agricultural entrepreneurs, as decision players for food and nutritional security at households and at community level. They also play crucial roles for of food supply stability in times of economic adversity. Farmwomen mostly work as sustenance farmers besides domestic activities, frequently go through poverty trap which poses challenges for their wellbeing. Although women farmer have massive labor involvement in this sector, women's contribution is frequently unseen and unappreciated.

Challenges like large climatic inconsistency and climate stress, together with limited investment in mountain-based agriculture and rural progress have resulted in least productivity and dearth of off-farm employment opportunities, forcing men into looking for alternate livelihood options elsewhere (Dar et al. 2017). As a result of migration, additional workload leaves farmwomen with less time to participate in capacity building program even if opportunities are available. Although new developments in agricultural divergence go along with commercialization and marketization creating chances for off-farm waged work, rural women with poorest education, insufficient training, and social immovability repeatedly face difficulties in responding confidently to these opportunities.

In perspective of feminization of farming, overall low female literacy has grim implications on future agricultural productivity and food security. Least educated rural women face the worst effects of structural changes in economy, mainly in agriculture sector that is oriented oward competitive international markets.

As a result, in mountainous region empowering women with better education will is a needed intervention and a valuable investment stregthening human capital for rural and agriculture growth leading to optimistic results for household and national food security mission.

Rural women continuously face severe discriminations in access to asset ownership, resources and agricultural support facilities. In addition, major portion of farmwomen work in farms is unpaid, making them often invisible. Farmwomen also tend to have very insignificant rights and access to natural resources, such as land and water. In order to attain sustainable and transformative changes, it is vital to address the current gender inequalities hindering farmwomen to make productive use of their potential, positively impacting food security and nutritional outcomes.

9.2 Addressing Work Load of Women: Issues, Solutions, and Way Forward

Rural women cover one quarter of global population and constitute 41% of global agricultural labor force, and this value rose to 49% for low-income countries (International Labour Office 2019). Earlier entire farming system in cold arid desert regions was dominated by cereal (wheat, barley, and oats) farming, which was reasonably inadequate for livelihood of tribal farming community of the region (Padma Dolkar 2018). These circumstances led to higher poverty and malnutrition in the region and inhabitants of the locality struggled much harder for their livelihood than people living in other parts of the country. Adverse situations led to higher migration to surrounding cities with higher vulnerability of poor people in remote regions.

Women support family business and works as farmer by performing all agricultural activities like ploughing of land, weeding, seed bed preparation, nursery raising, planting, harvesting, threshing, and post-harvest processing of crops grown as well as tending to small livestock and also serve to sell surplus production at local markets. Women also play dominant role in looking after family, firewood collection, fetching water, and cooking food. The multitasking role performed by women also leads to competing demands due to multitasking done by women. It adversely affects their quality of life and decision making. With additional workload they are unable to make the use of economic opportunities which cater to additional income generation.

At the household level we saw, inadequate consciousness and gratitude toward work done by women. This leads to power inequities within household whereby women finds it difficult to discuss their needs and support they require. Many a time while allocating the household resources, women needs are kept aside.

As far as farm mechanization is concerned, many a time women are left behind since the machines are not developed putting the needs of women at center.

Lack of capacity building programs, less access to education and resources are amongst the major obstacles leading to poor adoption and access of labor-saving technologies by women.

9.3 Biased Extension Services Ignoring Women Farmers at Large!

Women being less educated affect their capacity to rent or invest in new technologies. Extension and latest technical guidance is more pro toward men than women (Rathgeber 2011). Information on agricultural technologies usage is given by agricultural extension workers, major fraction of these master trainers are men who may not recognize the importance of women engagements as trainees. On most occasions extension guidance is accessible to men with pre-assumption that said information is passed on to farmwomen (Catherine et al. 2013). Women are more sensitive to risks in operating heavy agricultural farm machineries causing yield penalties and technological fade.

Women in rural areas are exclusively vulnerable to climate change influences like drought, connectivity cutoff, poor market access, lack of service center hubs and agricultural inputs because of factors including cultural norms and inequitable distribution of roles, resources, and power (Abid et al. 2018). A fair change to environmentally sustainable economies and societies unlocks important opportunities to strengthen women participation in decision making at all levels, ensuring sustainable livelihoods, improving working conditions, and advancing strategies for climate change. In majority of cases, it is found that the land ownership is frequently assigned to household head of the family, who are men (FAO 2018). Women therefore most frequently work on land possessed by their husband or the men of the households, thus leading to less access to new technologies. Services available also target asset owners, thus women trainings, information and technology requirements are not given due attention. It is therefore imperative to be conscious of dynamics in place between asset possession, tenure rights, and labor to ensure that women are able to adopt the technologies.

9.4 Gender Relevance to Viable Agriculture and Rural Growth

Gender equality is important in attaining food security, enhancing agricultural production, and improving lives of communities in rural regions, which are based on eradicating social and economic inequalities amongst men and women. There is equal need for intensifying social and economic possibilities for women. This is principally factual of improving women's reach to quality education, to provide firm platform for their part in decision making at household and society level. Economic independence expands bargaining status of women within household and helps them to raise their voice in decision making. Such empowerments in turn quicken exclusion of problems to increase their chances for self-employment and existence in labor market. Inequalities that farmwomen face not only decrease their contributions to agriculture by lessening their productivity, but also lead to have negative impact on wellbeing of families. If women gender gap is lessened or even removed, agricultural productivity will boost, which will ultimately result in less poverty, decrease in hunger and nutritional insecurity, and increase in economic development. Every so often rights and requirements of rural women stay inefficiently addressed or ignored at different platforms like law, national and indigenous policies, budgets, and investment policies at entire levels (Daley et al. 2013). Even when distinctive actions to improve rural women's situation are being planned, they are frequently not satisfactorily applied to mark actual change in lives of the rural women. Gender equality and rural women empowerment under the agenda "to leave no one behind" will benefit in attaining food security, refining nutrition needs, encouraging sustainable agricultural practices, and will help in evidence generation by stakeholders and decision makers. Thus, gender equality brings large contribution to nation's economic growth and it is a vital determining factor for food security.

9.5 Why Gender Equality Is Important in Agriculture

Despite women being major and primary farm producers, their roles are mostly either unrecognized or ignored (World Bank, FAO, IFAD 2009). Gender variances rising from socially built relationship amid women and men heavily disturb distribution of available resources among them and lead to countless disparities in progressive outcomes. Women are majorly involved as active traders, processors, laborers, and entrepreneurs, despite going through tough hurdles compared to men in accessing market. However, various development policies and projects accept men erroneously as key rural workers and farmers (World Bank 2007). Existing evidence specifies that distribution of land ownership is deeply tilted toward men (FAO 2018). Women especially face difficulty in making their voices heard even in democratic system and in enhancing responsibility and governance improvements in various areas.

9.6 Farmwomen Role in Food and Nutritional Security

Despite greater economic growth, women's poor status and less power to take decisions in South Asia is an important reason for Asian enigma of higher under nutritional rates than in sub-Saharan Africa. Agricultural innovations most possibly affect nutritional outcomes when involved in diverse processes and plans that readdress emphasis on agriculture food production and wider consideration of livelihoods, empowerment of women, and ideal intra household resources use. At times of crisis and conflict food grant and agricultural support are vital constituents of active interventions. Both these factors affect nutrition, food security, and livelihoods. Women are frequently forced to decrease their consumption in favor of other household members mostly males, which causes increased occurrences of malnutrition in women.

Levels of hunger and malnutrition are much higher in areas with a larger degree of gender inequality. Food insecure households are at larger risk of under nutrition, with adolescent girls, women of reproductive age, and young children being vulnerable to risk because of their higher nutrient needs. Solid evidence shows that improving women's socioeconomic status always results in positive nutritional outcomes, mainly when combined with nutritional education. Among all dimensions of women empowerment, improving women access to nutrition, control over available resources and reducing women working time and labor constraints seems to be more cumulative.

9.7 Importance of Gender in Land Policy and Supervision

Gender is an elementary determining factor of social relations in households and rural societies. Major factors including class, origin, and caste, gender defines to a great range a person's opportunity, aspirations, standard of living, access to resources, eminence in community, and self-perception. In addition women's right to resources impacts their capacity to produce and their conduct as producers. Depending on the norms leading household decision, pooling of income the women may not completely contribute to these benefits if it does not pertain to self-determining or direct rights on household land. Rights to natural resources and land raise women bargaining influence within household and societies, which leads in allocation of household resources to women. Panda and Agarwal (2005) revealed that women having property ownership are least susceptible to domestic violence in certain regions of India. However, cautious program designing, preparation, and execution are required because probable responses to empowerment procedure are domestic violence and community retaliations to women who look for self-governing rights. Achieving gender equity with respect to land rights as a result is determined not only by legal recognition of availed rights but also on disabling social and cultural restrictions. Regulations pertaining to implementing formal land and property in ways that deal gender bias regarding land rights, land access, permissible education programs, gender trainings for implementers must lead by women in formulating, planning, and executing programs at ground.

9.8 Agricultural Extension Services, Trainings, and Capacity Development Programs for Farmwomen

The different kinds of agricultural producers extending from very small private farms to agribusinesses have varying technical requirements and knowledge gaps that are not essentially met by present services. In specific huge number of rural households having very small land holdings have a tendency to produce restricted number of commodities for resident markets. They require extension services focused on usage of new technological interventions, marketing, and loan processing. The sex-disaggregated data about the entrepreneurs in rural areas are limited, as a result understanding the women's needs as business processors and executives in entrepreneurshipeurship remains a challenge. Engagement of women as entrepreneurs is dominant amid smallest sized enterprises or in individual possession. Women are likely to have businesses setup in saturated markets, small size enterprise and return limits restrict them to invest heavily. The women in business often face isolated in professional associations, financial institutions inclusions, chances for business associated trainings and market linkages.

9.9 Capacity Development for Small-Scale Women Entrepreneurs

Suitable interventions for capacity development can be planned only if specific primary vision has been achieved into overall opportunities and restrictions that women entrepreneurs and producers normally face.

This vision must be advanced through examination of gendered restraints and opportunities in planned project location. A woman needs to be correctly well versed about numerous business choices. Potential markets and their profitability must be considered in various credit and skills development programs carried for women. In formulated projects loan formalities and loan items must be suitable to necessities of poor women sections for successful outreach. The elevation of income making events for women mostly needs much more practical tactic than is frequently accepted by development programs.

9.10 Identifying and Characterizing Women's Enterprises

Women from their own home can manage microenterprises and face disadvantage of larger risk repugnance, local co-ordination and lower ability to fit into prescribed and detached market. In few cases women do manage numerous microenterprises concurrently to spread threat or conceal accurate magnitude of their earnings from men companions.

9.11 Identifying and Addressing Skill Gaps

Women entrepreneurs are involved in highly competitive global and domestic markets. Advanced expertise and knowledge are essential to move in major export markets such as proficiency in bureaucratic dealings, awareness of national standards, marketing channels, and consumer preferences. Women willing to move in export markets do require acquiring latest expertise to meet necessities that do not apply in the domestic markets. Although entry in global value chains is very demanding but once entry is achieved, extra learning may happen through supplier–buyer connections. Entry into worldwide value chains does have constructive influence on technological competence and upgrading skills (Humphrey 2004). Inclusive need-based quality training programs and effective partnerships are required to mainstream women filling and addressing the skill gaps.

In order to provide an equal opportunity to women enterprepreneurs, it is important to design and implement need-based capacity-building programs focusing and catering to their ecosystem. The design of the training frameworks needs to be tailored efficiently focusing on the financial literacy, and skill development being subjective to the need of the candidates.

9.12 Devising Inclusive Need Based Capacity Building Program

Implications of precise methods to capacity expansion are crucial but important.

Women instructors and women extension personnel can be more suitable because of cultural barriers that restrict interactions between men and women who are unknown to each other. There has to be interventions supporting comfortable travel to women instructors. Exposure visits allow entrepreneurs to have direct assessment of successful income generating activities and production techniques carried by exemplary entrepreneurs.

9.13 Labor-Saving and Women-Friendly Technology Evaluation

There is abundant indication that women offer considerable involvement to agricultural labor force as home care takers, sustenance farmers, entrepreneurs, and as daily wage workers. Thus, they need access to technologies that can ease up their work loads. Women work load not only disturbs their individual capacity to have choices but it also has a wider influence on the nutritional status and affluence of their families and societies.

While introducing the labor saving technology the crucial stages involve its effectvieness to lessen women work load. This can be measured from the discrepancies in the time taken to complete the activity by using the assigned technology and the time saved by adopting the technology.

How to encourage women's acceptance of labor-saving technologies and practices:

Reassuring gender sensitive policy atmosphere for technology development and dissemination can be one way, accepting as women have the same proprietorship and control over vital assets. Allowing them to plan ways for technology acceptance, collateral credit services and control over natural resources is important.

Gender sensitive technological expansion and dissemination policy. Policy encouraging communication exchanges between pertinent stakeholders analyzing labor and technology related gender data and further sharing the evidence of acceptance.

Women's capacity and autonomy to choose about technology requirements and use:

Still in most rural culture, uneven household dynamics still govern and women in spite of their work loads are not allowed to make choices, control their production desires, and let single-handedly control outlay of income they create.

9.14 Fostering Uptake of Labor-Saving Technologies Particularly for Tribal Farmwomen of Cold Arid Himalayan Deserts

There is strong evidence that women offer considerable involvement to agricultural workforce; however, their ability to adopt advanced technological interventions is very low. Outreach approaches that count on individual entrepreneurs frequently lay women and poor farmers at a disadvantage. In order to foster bigger production and value chain, preferment of technological interventions that can be adapted to diverse agro-ecological zones, climates and are able to reduce food losses, increasing access to markets is needed to retain youth farmwomen in agriculture. Choices to new and cheap agriculture machinery could possibly increase labor productivity, reduce drudgery and learning new skills to shift from subsistence to more market-oriented farming that can transform rural gender relations and lessen inequalities. Sustainable agricultural mechanization in tribal mountain areas must be tailored to the mountain diversity and farmwomen needs. There is a need to explore opportunities to introduce customized technologies ideal for women farmers in cold arid deserts. Farm mechanization pathway accepted by developed regions of nation only favored the plains and larger plot sizes and neglected needs of small farms in the hills and mountains. Alternative pathways to develop machines and tools suitable for sustainable growth of hill and mountain farming systems need to be confirmed. Advanced technologies for agriculture are majorly knowledge intensive and thus demand better practices to manage resources. It is important to assess benefits and hindrances of such innovative technologies for farmwomen of remote mountain areas. Mechanization hiring services through which most smallholder farmers particularly farmwomen can access machinery as per their requirements are to be focused. The majority of such services must come from either service center hubs or individual owners who rent them out. Formation of farm tool hubs by women resource pooling, and the income from tools rented out is used to purchase surplus tools so that farm tool hubs gradually expand. This process offers significant income gains and also advances access to agri-machinery to farmers who are incapable to afford them.

The major blockades in accessing and adopting labor-saving technologies include inadequate ability to access technology and its services, low education and literacy levels of farmwomen, major technologies are men purview, substantial liability of unpaid household accountabilities leads to time poverty of women, transport, and restrictions on mobility.

9.15 Recommendations for Sustainable Development of Himalayan Cold Arid Desert Agriculture to Improve Food Security and Nutrition

The major challenges faced by Himalayan cold arid desert agriculture include harsh climatic conditions, poor infrastructure, deficiency of services, and natural calamities which have terrible impacts on sustainability. Poverty, food deficiency, and nutritional insecurity are unattended challenges in remote areas affecting higher fraction of people. In these areas' inaccessibility, fragility leads to minimal economic opportunities, deprived market access, and poor institutional services.

Superior consideration and focused way out targeting the hilly areas are necessary to address poverty, food security, and nutritional supremacy. Mountains have acute role to provide ecosystem facilities for sustainable growth of downstream regions, but larger section of mountain population stays poor. Sustainable development of mountainous ecosystem and agriculture needs extensive investment in a holistic and cohesive approach. These approaches must involve policy framing, socioeconomic uplift programs, natural resource management, crop and livestock development. It is important to address these concerns considering the distinct circumstances of diverse farming structures.

Major policy, socioeconomic and institutional consideration for mountain sustainable development includes:

- Development and implementation of new strategies to address challenges being faced by different farming system from decades.
- Active and continuous consideration to risk supervision and frame policies for prevention, mitigation, and relief to manage natural disasters.
- Emphasize on betterment of existing institutes to offer public services comprising extension and microcredit to sustenance viable mountain agricultural expansion.
- Include councils of mountain groups in decision making, execution progressions of policies and take great initiatives to care mountain agricultural expansion.
- Increase occupancy to land holdings to guarantee extensive term investments.
- Focus to develop farmer enterprises to transmit subsistence agriculture to marketable agriculture for better productivity and revenue gains.
- Improve infrastructure of mountain regions to care agricultural growth and advance livelihoods.
- Enhance resilience to farming communities and small agro industrial enterprisers by marketing linkages and providing larger subsidies for value-added commodities to expand returns of mountain women involved in these enterprises.
- Launch mountain agriculture fund for renewal and support of agricultural amenities in these remote areas.

Sustaining natural resource a must scenario in mountains:

Encourage protection and viable use of available resources particularly water, biodiversity, soil, natural pastures. Ensure adaptation and mitigation measures to manage natural risks as an outcome of climate change and their gravity on natural resources. Avoid degradation of land and construct fertile soils through crop diversification, conservation agriculture techniques, promote agroforestry landscape model, water harvesting for efficient water management, control open grazing and conserving biodiversity.

9.16 Crop and Livestock Improvement Programs

Make sure progress of livestock, integrated crop, and rangeland farming types for efficient natural resource use. Boost farming of traditional crop germplasms and highlight key role of mountain-based farmers as protectors of these crop varieties. Create livestock community breeding programs where infertile animals are removed out to retain only productive animals with improved production. Diversify production systems and inspire farmers for cultivation of high value crops comprising fruit trees, vegetable, and value-added products to advance nutrition, income, and livelihoods. Sponsor protected agriculture to strengthen production of high value crops and improve water use efficacy and production effectiveness per unit areas in mountain-based agriculture systems.

9.17 Research and Capacity Expansion Programs

Improve long tenure investment programs for research in national institutes to enhance abilities to address challenges being faced by mountain agriculture for sustainable development. Sustenance capacity building initiatives and extension outreach to diverse stakeholders involved in mountain agriculture. Start farmer's schools and women empowerment drivers for transmission of technologies and skills development to address challenges faced by different stakeholders.

9.18 Regional and International Cooperation

Challenges and crisis being faced by mountain ecosystems are habitually complex and are tough to be fixed by single institution. Thus, it is vital to build up and improve cooperation at local and international levels.

The execution of mentioned recommendations will contribute highly to guarantee healthy mountain ecosystems with better agriculture production potential that will contribute to regions food security and nutrition and will also sustain available resources.

9.19 Group Farming (Putting Collective Farming Over Individual Farming)

Group or collective farming leads to sharing of input costs, risks, and profits. Harvest from group farming not only meets women household consumption requirements but also proves to generate income sources. Consolidation of minor productive divisions into superior units can meaningfully increase profits. Collectives permit women to share risks and encourage experimentation with new technological interventions and higher value crops, in addition gives chances to preserve certain joint belongings, like water, soil quality, and other common resources available. Economies of scale can be vital for packaging, transport, and marketing plans. Women collective farming provides power to group in making farming decision. Such interventions offer several avenues for landless women if lead by main farmers in community who share best practices in farming by sharing their farming experiences, credit linking strategies, indigenous knowledge, and information on the practices to promote environment friendly agricultural techniques. Critically group farming into collectives, cooperatives, unions always support their voice in taking decisions and in involvement to formulate and evaluate policies.

Supporting women rural producer groups to emerge as effective chain partners is often slow process. Thus, step-by-step development of capacity improvement may be essential to get them involved in more tasks. Post-production market connections need to be powerfully constructed seeing profit as bottommost line. Increasing women's access to off-farm engagement i will help in increased accessibility and better experience to improve their skills.

9.20 Gender and Livestock

Livestock offers employment opportunities, income generation, and better food and nutrition security through diverse production systems. Various challenges faced by livestock enterprise include guaranteeing food, resource and livelihood security to deprived smallholder producers and processors. Applying gender lens to recognize and address different requirements and limitations linked to pertinent livestock and value chains is vital for defining most ideal outcomes and most efficient use of available resources. Gendered roles and vision are not set in stone and are structured to transformation for diverse social, economic, environmental, and health associated aims. Transversely in regions, women are comprehensively recognized for their abilities in poultry, small ruminant, and micro livestock invention and dairying comprising processing and marketing of milk products and milk. Ever more, experience demonstrates (Bravo-Baumann 2000) that women role remains unrecognized and underappreciated in designed livestock policies. Their poor market access, services, technologies, information, and credit declines their capability to expand productivity and profit from livestock enterprise. But being an avenue of self-employment livestock boosted their integrity significantly.

Livestock revolution is a fundamental phenomenon that is here to stay. Technologies for increasing production will also give challenges to rural women. Its future in India is largely based on approaches followed. Policy deficits will not halt the livestock revolution but will certify that the form it takes is least promising for growth, poverty, reduction and sustainability. Women have important role in livestock production in rural areas, but their existence in remotest parts makes them least economic gainers. Thus, formulating new strategies is important to improve rural women livestock-based livelihood. Meat production and demand has increased as economies got strengthened and income levels have increased. Improving role of women in livestock production with access to inputs and services has the potential to decrease malnourished people across globe. Unfortunately, women lack access to investment as livestock farmers, thus there is a need for unbiased system that accounts for all individuals within communities. Mainly in low-income nations it is essential to know various roles and reasons for keeping livestock, among them dominant ones comprise food production, income generation, financial investment, social status, and manure in the context of vulnerable demographics. To overcome these concerns developing policies and strategies to prioritize investments in livestock sector is essential. Financial support, infrastructure, and modernization in marginalized and vulnerable demographics which majorly include youth and women need to be streamlined.

9.21 Gender Responsive Extension, Financial and Agricultural Research Services

Extension services need to get reformed to serve women better. A survey carried by FAO in 1988 and 1989 covering 97 countries came with outcome that about 5% of extension lead facilities were addressed to the rural women and merely 15% of extension advice-givers were female (FAO 1996). Only just 16 researchers from the World Bank and IFPRI recognized enormous gender disparities in accessing extension services in surveys completed in India. The failure of extension services to women farmers seems to be attributable to common assumption that extension services transferred to men spontaneously trickles down to women. Such assumptions often ignore the fact that the needs of women are different from those of men as far as the extension approach is concerned.

Poor farmers face hindrances in attaining crop insurance, predominantly for compensation after disastrous produce and in gain access to credit. As per 2003 rural finance access survey in India about 87% small or landless farmers had no access to institutional credit (Garikipati 2008). The difficulties are multiple and the infrastructure for economic amenities in rural regions is frequently weak or absent. For women such problems are of course exaggerated due to hindrances they face in recognition of their rights to land which make access to credit even tougher. Maximum viable institutions offer services to women in very smaller extents than to men. In our country India now, instructions are that a third of entire loans by state development banks essentially go to women.

Agricultural research and development is one area where inclusion is weak while developing new varieties, livestock, and introduction of new technological interventions. As a substitute, agricultural research and expansion must take many constraints into account faced by women to efficiently alleviate poverty and empower women. Complications in accessing credit is also an issue for women. Reorienting agricultural research to be gender friendly will create platform where both men and women can play different roles in production and marketing practice, access to productive resources, adopting new technologies, depiction of female and male scientists, and extension workers in agricultural extension and research systems. Data collection and evidence generation needs to be inclusive.

9.22 Institutionalizing the Effort for Equity Through Krishi Vigyan Kendra in Himalayas for Mountain Women Farmers

The economy of cold arid Himalayan tribes continues to be predominately based on agriculture with small landholding surrounded by glaciers & wastelands with least irrigated facilities. These communities have higher incidence of illiteracy, malnourishment, abject poverty faced social marginalization and geographic isolation. KVK Kargil II (Zanskar) is striving hard for livelihood security and socioeconomic upliftment of the tribal farming community of region. The challenges in the region are manifold, which include extremely harsh and long winters, short crop growing season, poor road connectivity, poor communication facilities, poor marketing structures, centuries of isolation and socioeconomic marginality. The station, right from the beginning has adopted a multipronged and pragmatic approach to deal with these challenges. In tribal subplan program brief account of achievements made is presented in form of success stories. The project "Technology interventions for improving tribal farmer's food, fodder, income and livelihood security" was funded by ICAR and led by Dr. Shahnawaz Rasool Dar with objectives: (I) Eliminate hunger, food insecurity & malnutrition by introduction of improved & high yielding varieties to improve food, fodder, and nutritional security of tribal farming community. (II) Making Cereal & vegetable production system more productive & sustainable by promoting resource conservation agriculture technologies & protected cultivation in vegetables. (III) Organizing animal, soil and plant health camps, farmer participatory cluster demonstration field days to address location specific problems and prospectus of agricultural interventions among tribal community of cold arid deserts of Himalayan region. (IV) Make livestock more productive with major focus on improving animal health by concentrate feeding, disease management, and value addition in different fodder crops. (V) Reduce rural poverty by providing skill development to tribal women folk and rural youth across multiple agricultural related enterprises like food processing, knitting, tailoring, pashmina & wool processing, mushroom cultivation, vermincompost preparation, etc. (VI) Promoting farm mechanization to reduce women drudgery and creating business opportunities for local youth as service providers to bridge yield, livelihood gaps and social equity, and (VII) Transforming Ladakh agriculture from sustenance level to commercial level (Table 9.1).

9.23 Transformative Approach Case 1: Krishi Vigyan Kendra Kargil-II (Zanskar) Harvest Record Wheat Yield Under Cold Arid High-Altitude Mountains of Zanskar

Agriculture in Zanskar is confronted with multifaceted constraints and herculean challenges. Despite immense toil and hard work of the farming community, the farm returns so far have only been dismal. Extremely harsh winters, poor soil health, unavailability of quality inputs, geographic remoteness, poor road connectivity, small land holdings, and lack of scientific know-how are some important factors that make profitable crop production an extremely daunting task in this region. Krishi Vigyan Kendra Kargil-II Zanskar, in its endeavor to alleviate the distress of the local farming community, focused on prioritizing the pressing issues which need immediate attention. One such issue prioritized by the Kendra was identification of

Table 9.1 Head wise budget allocation to empower tribes of cold arid Himalayan deserts under ICAR TSP scheme

	Budget allocation			Proposed beneficiaries' coverage		
Item procured	2020–2021	2019–2020	2018–2019	2020–2021	2019–2020	2018–2019
Farm tool kit	2,640,000	840,000	4,300,000	463	306	194
UV films GHP small, Black Mulch & Others	1,885,000	5,422,000	4,599,000	1039	1981	1845
Animal feed	300,000	650,000	1,548,000	145	320	645
V compost & fertilizers	300,000	99,000	1,357,500	167	237	107
Poultry chicks	200,000	195,000	0	147	110	0
Seed & planting material	1,500,000	4,949,911	2,192,471	430	656	245
Animal medicine	300,000	600,000	800,000	434	670	732
Pasture development inputs	400,000	1,000,000	0	23	34	0
Trainings	0	200,000	150,000	0	23	13
Pashmina goats and sheep	0	5,000,000	410,535	0	200	40
Infrastructure (tribal hostel, animal clinic, compost units, bore well), POL & Others	13,272,500	800,000	692,739	Community based developments		
Total	20,797,500	20,490,000	16,050,245	2848	4537	3821

high yielding wheat varieties feasible for the region. The Kendra has been evaluating a wide range of crop varieties through on farm trials (OFTs) and front-line demonstrations (FLDs) under variable climatic ecologies across the length and breadth of region. One of our most striking stories of 2021 from Zanskar is that of an OFT laid out by Kendra at the fields of tribal farmwomen group involved in group farming and participatory wheat seed production in village Tungri had harvest record wheat yield of 5.8 t ha^{-1}, the highest ever recorded in the history of the Zanskar region.

At the time of harvesting, the Kendra organized a field day & invited farmer particularly women farmers of surrounding villages and area-specific partners to show them the actual yield that was obtained under high yielding varieties with best agronomic managements. Hon'ble Chief Executive Councilor (CEC), LAHDC, Kargil & Deputy Commissioner, Kargil also visited this wheat field prior to harvesting and appreciated efforts of Kendra in transforming agriculture to secure livelihood of tribal farming community of remotest region. Dr. Shahnawaz Rasool Dar, Head KVK Kargil II Zanskar said that the aim of organizing this event was to push the promising technologies to other farmers and policy makers of the region.

High yielding varieties of wheat demonstrated in farmer participatory seed production mode seems to have created a much-needed confidence among the local farming community. Based on performance of introduced varieties at the event, about 50 farmers requested for the seed of these high yielding promising varieties. Involved farmwomen group leader said that such changes with improved yields, have much to offer in terms of nutritional and livelihood security of tribal farming community of high altitude of Himalayan region. Among the tested varieties DBW 187 provided highest yield of 6.1 t ha^{-1} followed by HD 2967 which gave yield of 5.8 t ha^{-1} HD 2733 yielded 5.3 t ha^{-1} against 3.5 t ha^{-1} of the decade's old local grown Kromar and Krokar wheat varieties, which presently covers about the 80% varietal spectrum of the region. Given the low productivity base of the region, the yield gain of this magnitude is highly remarkable. SKUAST-Kashmir under tribal subplan scheme is firmly committed to revive fortunes of the tribal farming community through transformation of local agriculture on modern scientific lines and capacity building of farmers. Plugging big holes in the existing agronomy could have a big impact on the profit margins in this area. The way these high yielding varieties delivered in terms of record wheat yield, cold arid deserts of Ladakh have a tremendous scope to amplify the system productivity from the existing levels.

9.24 Transformative Approach Case 2: Livelihood Security of Tribal Women Farming Community of Cold Arid Himalayan Deserts by Improving Fodder Production and Livestock Health and Production

Zanskar is a resource poor, landlocked valley in Kargil district of Ladakh. Livestock population of the region is high and constitutes the most important enterprise of the tribal farming community. However, shortage of quality animal feeds and fodders and poor scientific know-how severely constrains the production, reproductive and overall health performance of the livestock.

In order to alleviate the problem, KVK Kargil-II (Zanskar) adopted a multi-pronged strategy of replacing the low performing, non-descript poor quality local fodder varieties with high performing fodder varieties of oats and fortification of straw and various other plant by-products with urea and molasses treatment. The Kendra initially adopted certain villages in three blocks of the region as model fodder villages. All the inputs were provided to the selected women farmers under TSP scheme. Field days were conducted and progressive farmers from all across the Zanskar valley were invited to inspire them for change. A large number of awareness and training camps were conducted wherein women and men farmers were trained and sensitized about importance of good quality seeds in fodder production, conservation techniques, fortification of low quality fodders by urea molasses treatment, importance of balanced feeding, use of unconventional fodders for higher economic returns, rearing of livestock on scientific lines, adoption of anti-helminthic dosing regimens and importance of calcium and vitamin feeding supplements in dairy animals, etc. In the following years the progressive farmers were encouraged to procure the high performing fodder seed from neighboring model villages and popularize them in their respective areas. Today most of the local poor quality fodder varieties and traditional method of animal rearing are replaced with high performing fodder varieties and rearing of animals on modern scientific lines, respectively.

The interventions have significantly increased the yield and quality of fodder (Fig. 9.1) and led to multifaceted improvements in the production, reproductive and overall health performance of the livestock. General body health, weight gain, milk yield, quality and quantity of wool in case of sheep, live birth weight of young ones, and conception rates have significantly increased (Table 9.2). Drastic reduction in infant mortality rates and post-partum disease occurrences in dams with increased survivability of neonatal calves, kids, and lambs have greatly increased economic returns of the tribal farmers. Efficient management of livestock in resource poor region of Ladakh has been one of the most important achievements of the Kendra to safeguard the livelihood security and overall wellbeing of the tribal farming community of cold arid Himalayan deserts.

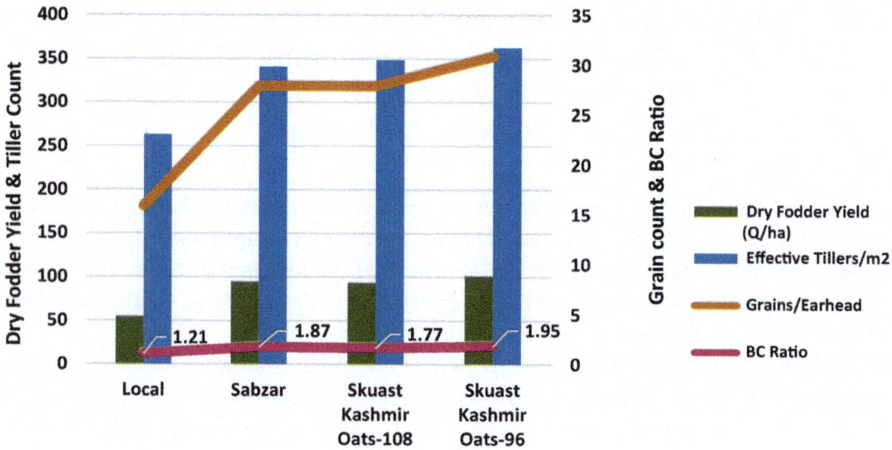

Fig. 9.1 Oats varietal performance in cold arid Himalayan deserts of Zanskar. (Average of 217 demonstrations)

Table 9.2 Effect of different feeding standards on livestock production & reproduction performances of different animals in Ladakh

Treatments	Sheep and goat (average of 145 each)			Dairy cattle (average of 107)		
	Wool yield (g) / animal	Body weight (kg)/ animal	BC ratio	Average milk yield (Lt)/day/ animal	Average calving interval (months)	BC ratio
Local fodder feeding practice	650	23	2.23	2.1	19	1.7
Oats, Fortified Fodder Crops & Scientific Rearing	950	30	2.81	3.3	13	2.7

Farmwomen Feedback on Fodder Interventions The newly introduced fodder varieties were excellent in production, better in acceptability and palatability by animals, and more resistant to diseases and pests as compared to local varieties. A large number of farmers have not only become self-sufficient but also sell their surplus produce in local market, which was unheard of before the intervention. Fortification of the straw and other farm by-products ensures availability of cheap and quality dry feeding material during six-month long winters when area remains cut off from the rest of the world due to heavy snow fall.

9.25 Transformative Approach Case 3: Role of Youth in Breaking the Yield Barriers in Vegetable Production and Entering in Seedling Business as Entrepreneurs (See Table 9.3)

Table 9.3 UBRAK women SHG LED nursery raising enterprise: where aspirations are translated into action

No. of members in group: 6 Total area under nursery (0.1 ha)				
Production performance at nursery level				
	No. of seedlings sold	**Gross return INR**	**Production cost INR**	**Net return INR**
Onion (Red Corel)	50,000	75,000	25,000	50,000
Cabbage (Mitra)	40,000	60,000	18,000	42,000
Total	90,000	135,000	43,000	92,000
Production performance at farmer's field				
Name of varieties	Yield (q/ha)	Bulb or head weight (kg)	B C ratio	
Onion (Red Corel)	134.2	0.16	2.78	
Local	74.5	0.09	1.64	
Cabbage (Mitra)	340.1	0.94	2.62	
Local	193	0.67	1.49	

Non-availability of quality planting material is one of the main problems for profitable vegetable production in Zanskar. The Kendra adopted Ubrak women self-help group comprising of six progressive farmwomen for growing of quality nurseries of onion and cabbage under protected conditions. The SHG was provided with 500 g of onion seed (Red Corel) and 400 g of cabbage seed (Mitra) under TSP scheme. All the operations from sowing till uprooting were closely supervised by the Kendra. A total of 50,000 and 40,000 seedlings were harvested from the onion and cabbage nursery, respectively. The seedlings were sold to the neighboring farmers @ Rs.1.5/seedling. Net profit obtained from onion and cabbage nurseries was Rs.50,000 and Rs.42,000, respectively. Total net return per member: 15333 INR. Production performance of sold out seedlings was evaluated at farmers' fields. Yield as well as benefit cost ratio of Red Corel and Mitra was significantly higher than local varieties.

9.26 Transformative Approach Case 4: Krishi Vigyan Kendra Kargil-II Organized 1 Day Kisan Mela on Nutritional Security

Krishi Vigyan Kendra Kargil-II (Zanskar) celebrated 1 day Kisan Mela at its main campus Padum Zanskar on 23-09-2021. The key theme of the Mela was "Empowering Tribal Youth & Women for Technology Driven Farming in Cold Arid Deserts of Himalaya." Farmers from different villages, officers of the line departments, BDC chairmen, Sarpanches, Panches & other local representatives of Zanskar region participated in the event. Fresh farm produce & value-added products by Krishi Vigyan Kendra groomed progressive farmers & Self-Help Groups (SHGs) were showcased in the Mela. Moreover, advanced farm machinery, high quality local processed products, seeds of promising high yielding varieties of cereals, fodders & vegetables, and other latest technologies introduced by the Kendra were also demonstrated. KVK stressed upon the importance of adopting modern scientific tools & technologies, high yielding varieties, crop diversification, quality fodder & mineral supplements for improvement of animal health and conservation of natural resource base. Scientists of SKUAST-K for the welfare of tribal farming community of Ladakh region highlighted role of Kendra under tribal scheme program with theme "no one should die of hunger." Scientific team of Krishi Vigyan Kendra expressed satisfaction that the interventions made by the Kendra during the past few years have started to bear fruits at ground level. The officers of line departments laid special emphasis on organic farming & also apprised the participants about various government schemes available for the welfare of farming community of Ladakh. Top 11 performing farmers were presented awards and consolation prizes were also distributed among the remaining participants who displayed their produce during Kisan Mela. Introducing the tribal community to vegetable cultivation helped to reshape the village economy and aided in the overall empowerment of tribal women.

9.27 Transformative Approach Case 5: Integrated Farming Systems Drivers for Food and Nutrition Security of Trial Population

Food and nutrition insecurity is a persistent problem in almost all the tribal areas. To combat these issues, integrated farming system involving home nutria-gardens with back yard poultry rearing and nutritionally fortified dried cheese locally known as churpay was introduced. This intervention not only helped women to eradicate problems related to nutritional deficiency among the women and children, but also gave them opportunity to acquire nutritional security and sustainable development. In IFS demonstrations laid, nutri-gardens provided vegetables for family consumptions and vegetable wastes were used as feed for the backyard poultry rearing. Improved varieties of chicken like Keystone for egg laying and dual purpose variety of Vanraja were introduced for the very first time in Zanskar valley and their performance and impact showed promising results. Modifications related to local housing and feeding management of backyard poultry helped in yearlong survival in extreme hard and cold conditions where temperature dips to −30 °C during January & February. On an average around 175 eggs per year per bird and meat of 4.0 kg per bird were recorded. Cost–benefit analysis showed higher indicators of 1:5, meat and eggs produced where purely organic in nature. High protein diet is essential component for the people of cold regions during winters and it can be successfully achieved through strengthening the integrated backyard poultry system in these remote areas. A simultaneous component in integrating components of farming system involves strengthening of local dairy product preparation and processing techniques on modified scientific lines so as to improve the keeping quality and nutrient value of the farm produce. Interventions through capacity building programs in the cheese making on scientific techniques and its drying with introduction of solar driers with least possible contaminations during processing to make churpay of high quality were achieved. Churpay are cooked during winters and act as high quality foods for pregnant women and children. Moreover, for efficient and quality butter production with tabletop curd churner not only helped in processing of extra milk produce but also provide a hope in dairy sector entrepreneur opportunities for rural youth. One such example is cheese and butter production at commercial level by processing of extra milk produced during peak season in the village Atting. The farmer Ishay was recognized at National level during Northern India Regional Agriculture Fair 2020–2021 for producing quality organic butter and cheese and same time played vital role by retaining youth in Agriculture as group entrepreneurs to create employment opportunities to tribal women. Integrating the various components of the tribal farming systems was not only attractive to women but it also provided both nutritional security and economic viability and returns.

9.28 Conclusions and Recommendations

Women farmers in agriculture while performing their roles due to less access to improved knowledge face massive challanges. Prevailing laws and rules particularly related to property and inheritance which differentiate between women and men need to be scrutinized. In order to empower women in agriculture it is important that they have land ownership and decision making power to improve their land productivity. Systematic presence of gender dimensions should be incorporated in all rural development projects with aim to have speedy progress toward food and nutrition security maximizing the pathways eradicating hunger and malnutrition. Women-based organization and pressure groups can play crucial role.

Very little is known and documented about social dimensions of technology access and acceptance in the mountainous region. Research programs on social value of technology and women will help adopting effective technologies over others. Consistent sex disaggregated data on time use needs to be collected and analyzed for future policy making and framing to design new projects and programs. Research in agro-processing technologies must be carried out to reduce time use and increase income opportunity for women entrepreneurs. Listening to women farmers is important.

Transport sector in tribal areas need to be strengthened to have safe and affordable transport for women farmers to improve their market access and also enable them to access other services. Women participation with cooperatives and producer organizations may facilitate access to markets in tribal areas. A participatory meeting involving both men and women needs to be organized sharing the concern of each other in improving their agriculture practices. Projects need to increase their outreach to pledge gender consciousness through capacity development. Gender differences in adopting technologies need to be measured along with outcomes and impacts for securing livelihoods of tribal communities.

Major players involved in implementing gender supportive approaches and technologies to address women needs should include advocacy and targeted capacity development on gender equality programs. Training and capacity development of tribal farmwomen is essential to endorse their leadership and negotiation skills, support voicing of their concerns, and improve their technical skills.

Attention needs to be paid to increasing representation of female extension workers, specific targeting of women farmers, increasing knowledge of the technologies and extending outreach. In order to achieve this all resource persons need to be identified. There is a need to include more and more rural women in worldwide in the process of economic integration and commercialization of agriculture sector. There is an ever increasing need to empower rural women with capability to function efficiently using innovative technologies in agriculture, information, and communication.

Photographic representation of few activities carried with tribal farmwomen for nutritional security and their empowerment:

ZANSKAR
DISTRIBUTATION OF AGRICULTURE

FIELD DAY

Source: Photographs were taken by team KVK Zanskar

References

Abid Z, Abid M, Zafar Q, Mehmood S (2018) Detrimental effects of climate change on women. Earth Syst Environ. https://doi.org/10.1007/s41748-018-0063-9

Bravo-Baumann H (2000) Gender and livestock: capitalisation of experiences on livestock projects and gender. Working document. Swiss Agency for Development and Cooperation, Bern

Catherine R, Berhane G, Tadesse F, Taffesse AS (2013) Gender differences in access to extension services and agricultural productivity. J Agric Educ Ext 19(5):2013

Daley E, Flower C, Miggiano L, Pallas S (2013) Women's land rights and gender justice in land governance: pillars in the promotion and protection of women's human rights in rural areas. Int Land Coalition

Dar SR, Norphel C, Akhoon MM, Zargar KA, Nazeer Ahmed MA, Yabgo KA, Dar NH, Thomas T, Singh M, Kumar A, Hussain S, Kumar B, Baba AY (2017) Man's artificial glacier—a way forward toward water harvesting for pre and post sowing irrigation to facilitate early sowing of wheat in cold arid Himalayan deserts of Ladakh. Renew Agric Food Syst. https://doi.org/10.1017/S1742170517000527

Dolkar P (2018) An overview of transition in traditional agriculture of Ladakh. J Himalayan Ecol Sustain Dev 13(2018) ISSN 0973-7502:26

Food Agriculture Organisation (1996) FAO focus: women and food security: women hold the key to food security. Rome

Food Agriculture Organisation (2018) The gender gap in land rights. Research programme on policies, institutions and markets

Garikipati S (2008) The impact of lending to women on household vulnerability and women's empowerment: evidence from India. World Dev 36(12):2620–2642

Humphrey J (2004) Upgrading in global value chains. Working paper no. 28, policy integration department, world commission on the social dimension of globalization, International Labour Office, Geneva

International Labour Office (2019) Empowering women in the rural economy, decent work in the rural economy policy guidance notes. https://www.ilo.org/global/topics/economic-and-social-development/rural-development/WCMS_436223/lang%2D%2Den/index.htm

Panda P, Agarwal B (2005) Marital violence, human development and women's property status in India. World Dev 33(5):823–850

Rathgeber EM (2011) Rural women's access to science and technology in the context of natural resource management. Institute of Women's Studies University of Ottawa. EGM/RW/2011/EP.6 September 2011

World Bank (2007) Agriculture for development: the gender dimensions. Agriculture for development policy brief. World Bank, Washington

World Bank; Food and Agriculture Organization; International Fund for Agricultural Development (2009) Gender in agriculture sourcebook. Agriculture and rural development. World Bank. © World Bank, Washington. https://openknowledge.worldbank.org/handle/10986/6603 License: CC BY 3.0 IGO

World Economic & Social Survey (2011) The great green technological transformation, Department of Economic and Social Affairs. E/2011/50/Rev. 1 ST/ESA/333

Chapter 10
Extent of Participation of Farm Women in Decision Making Regarding Agricultural Activities

Anuradha Ranjan Kumari, Barun, S. K. Mandal, and R. K. Mandal

Abstract Women's participation in the decision-making process has a significant impact on their improved status and greater role in the society. Their participation is potentially important to bring equality between women and men in order to achieve sustainable development. Rural Indian women are extensively involved in agricultural activities. However, the nature and extent of their involvement differs with the variations in agro-production systems. The mode of female participation in agricultural production varies with the land owning status of farm households. In overall farm production, women's average contribution is estimated at 55–66% of the total labor with percentages much higher in certain regions. Hence, the present study was attempted to ascertain the extent of participation and decision-making pattern of farm women in agricultural activities. The present study was carried out in Siwan district of Bihar by selecting 150 farm women. The results show that the higher percentage of farm women were of middle age group, illiterate, medium family size, marginal farmers, membership in social participatory groups, low material possession, low economic motivation, i.e. attributed to lower incomes. Further regarding psychological characters shows that majority of farm women were having 68 percent medium level of extension contract. Maximum 78.80% of respondents had medium role in decision making followed by high and low. The study revealed that more number of farm women found to have high level of participation in agricultural extension (45.17%) followed by the medium (33.84%) and low participation (20.98%) in agricultural operations. The study further revealed that a high number

The original version of this chapter has been revised; authors' affiliation has been updated. The correction to this chapter is available at https://doi.org/10.1007/978-981-19-6978-2_13

A. R. Kumari (✉) · Barun · S. K. Mandal · R. K. Mandal
Krishi Vigyan Kendra (DrRPCAU) Bhagwanpur Hat, Siwan, Bihar, India

S. Munshi, M. Singh (eds.), *Women Farmers: Unheard Being Heard*, Sustainable Agriculture and Food Security, https://doi.org/10.1007/978-981-19-6978-2_10

of farm women (52.07%) were observed in low category of decision making which was followed by (27.84%) medium and (20.07%) high categories of decision making. This is because of male dominance, no knowledge about improved technology, and lack of education. Majority of farm women shared the constraints of lack of time management between farm work and daily home activities as the real challenge. Most of the time management of household chores turned out to be another constraint faced by farm women.

Keywords Extent of participation · Agricultural activities · Decision making · Farm women

10.1 Introduction

Women play a pivotal role in agricultural and rural economics in all developing countries (FAO 2011). Women farmers account for more than quarter of the world's population. In Indian Society women have a multi-dimensional role. The largest numbers of women in India are engaged in farming operations either as cultivators or as supervisors or as agricultural laborers. Doss et al. (2011) clearly showed that women comprise about 43 percent of the agricultural labor force globally. They are the main participants in various agricultural operations like Seed sowing, Transplanting, Weeding, Harvesting, Threshing, Application of manure, Storage of seeds and Food grains, and Post-harvest home level processing (Chayal and Dhaka 2011). Rural Indian women are extensively involved in agricultural activities. However, the nature and extent of their involvement differs with the variations in agro-production systems. The mode of female participants in agricultural production varies with the land-owning status of farm household. Their roles range from managers to landless laborers. In overall farm production, women's average contribution is estimated at 55–66% of the total labor with percentage much higher in certain regions.

In the vegetable growers, women perform a variety of tasks both in cultivation and marketing. With the advent of science and technology, the rural society is witnessing unprecedented changes in all spheres of life. As a result of which it is found that negotiations regarding prices are made with the members of families involved in vegetable cultivation. Practices related to vegetable cultivation have now been considered as a family enterprise in which husband and wife participate to share work and have quality time to enjoy with each other. Thus, it is expected that all decisions related to practice of vegetable cultivation are also taken jointly.

Although women's participation in the decision-making process has a significant impact on their improved status and greater role in society (Begum 2002), their involvement in decision-making process specially related to money matters is low (Raju and Rani 1991). Gender equality in democratic governance is very uneven; in most of the world, women are under-represented in positions of power (Mumtaz and Aysha 1982; Slovenia 1998; Habib 2000; FAO 2003; Rahman 2008). In rural families, type and size of the family, caste, size of land holding, socio-economic status of the families, education level of rural women, their employment status and rational position affect their involvement in decision making. Illiteracy, poverty, and unemployment are the major problems of many developing countries, of which India is no exception. The growing problem of poverty in our country has promoted the economic planners to come up with various programs to curb poverty. Keeping the above facts in mind, the present study was conducted to find out the participation and decision making of rural households in various farming activities.

10.2 Materials and Methods

The present study was conducted in Siwan district of Bihar. Three blocks—Bhagwanpur Hat, Maharajganj and Badhariya—were selected for the study. From each respective blocks two village Bhagwanpur Hat, Sondhani, Kheduchapara, Sankarpur, Kailgarh and Badhariya were selected purposively. Among each village 25 farm women were selected randomly, resulting into a total sample size of 150 rural women. The data were collected through a structured and pre-tested interview schedule. The researcher personally approached the respondents and explained to them about the purpose of the study. After establishing rapport with the respondents, the farm women were interviewed and their responses were recorded in the interview schedule. The scale consists of 17 statements of agricultural operations and decision-making process. The responses of respondents were obtained in 3-point continuum, namely low, medium, and high. The scoring was assigned in the order of 1, 2, and 3, respectively. After establishing rapport with the respondents, the farm women were interviewed, and their responses were recorded in the interview schedule. The data were collected, tabulated, and analyzed to find out the findings and drawing of the conclusion. The statistical tools like frequency, percentage, and rank were employed to analyze the data. The present study was analyzed with the help of methodology explained by Hedges (1963).

10.3 Results and Discussion

10.3.1 Socio-Economic Characteristic of Farm Women

The data indicated in Table 10.1 explain the profile of farm women. As far as socio-economic and personal profile of farm women is considered the results shows that higher percentage of farm women were of middle age group (62.00%), Illiterate (47.33%), Medium family size (62.66%), Marginal farmers (64.00%), Membership in social participation (56.66%), Low material possession (59.33%), and Low economic motivation (57.33%). Further regarding psychological characteristics table shows that majority of the rural women were having medium innovative proneness (78.00%), medium scientific orientation (62.00%), and medium market

Table 10.1 Socio-economic and psychological characteristics of farm women

Characteristics	Response category	Frequency	Percentage $N = 150$
Age	Young (up to 30 years)	23	15.33
	Middle (30–50 years)	93	62.00
	Old (above 50 years)	34	22.66
Education	Illiterate	71	47.33
	Only read & write	24	16.00
	Primary	20	13.33
	Middle	19	12.66
	High school	8	5.33
	Higher secondary	8	5.3
	Graduation & above	0	00
Family size	Small family (up to 4 members)	20	13.33
	Medium family (4–8 member)	94	62.66
	Large family (above 8 member)	36	24.00
Land holding	Marginal farmers (up to 2.5 ha)	96	64.00
	Small farmers (2.5–5 ha)	30	20.00
	Medium farmers (5–7.5 ha)	24	16.00
	Large farmers (above 7.5 ha)	0	00
Social participation	Member	85	56.66
	Office bearer	24	16.00
	Non-office bearer	41	27.33
Material possession	Low (up to 4 score)	89	59.33
	Medium (4–7 score)	41	27.33
	High (above 7 score)	20	13.33
Economic motivation	Low	86	57.33
	Medium	39	26.00
	High	25	16.66
Innovative proneness	Low	12	8.00
	Medium	117	78.00
	High	21	14.00
Scientific orientation	Low	41	27.33
	Medium	93	62.00
	High	16	10.66
Market orientation	Low	21	14.00
	Medium	99	66.00
	High	30	20.00
Mass media exposure	Low	87	58.00
	Medium	40	26.66
	High	23	15.33
Extension contact	Low	17	11.33
	Medium	102	68.00
	High	31	20.66

orientation (66.00%). Regarding the distribution of respondents according to their mass media exposure and extension contact the table shows that out of 150 farm women, 52.00 percent had low level of mass media exposure and 68.00 percent respondents had medium level of extension contact.

10.3.2 Role of Farm Women in Decision-Making Process

The variable was presented in a four-point scale as most often, often, sometimes often, and never. A score 0 was assigned if she is never involved in decision-making process, score 1 was given when she is rarely taking decision, a score 2 was given if she takes decision often, and Score 3 was given when she is involved very much in decision making. Decision-making score is the total of the all scores obtained on different items of agricultural operations performed by farm women. According to range of scores, the respondents have been categorized into three groups as low, medium, and high. The data indicated in Table 10.2 explains the distribution of respondents according to their role in decision making. It is clear from Table 10.2 that out of the total 150 respondents, maximum 78.00 percent of respondents had medium role in decision making followed by 12.66 percent had high and only 9.33 percent had less role in decision making.

10.3.3 Extent of Participation Pattern of Farm Women

It is clear from Table 10.3 that more number of farm women were found to have overall high level of participation in agricultural activities, i.e. 45.17 percent followed by medium participation with 33.84 percent and low participation of 20.98 percent. Further data revealed that the farm activities in which farm women obtained the highest score at high level of participation were weeding (mean score 4.22), Selection of seed variety (mean score 4.60), harvesting (mean score 4.16), Soil testing (means score 3.38), winnowing process (mean score 4.43), seed treatment (mean score 3.69), plant protection (mean score 3.92), sowing (mean score 4.40), manure and fertilizer application (means score 2.52), Threshing process (mean score 2.50), seed processing (mean score 4.84), grain storage (mean score 4.55), Irrigation management (mean score 2.79), collection of harvested crops (means score 4.27), preparation of land (mean score 3.12), marketing (means score 3.52), and soil treatment (mean score 2.64). The study of Ghosh (2000), Badiger and Huilgar (2004), Sharma and Badodia (2016), Singh et al. (2005) and Yadav et al. (2005) also found higher participation of farm women in agricultural activities in this respect in the present study in conformity with these findings.

The perusal of the data presented in Table 10.4 revealed that the farm women participated in decision-making process in each and every farm activities. The data indicated that the higher number of farm women (52.07%) were observed in low category of decision making followed by medium (27.84%) and high category of

Table 10.2 Distribution of farm women according to their role in decision making

Categories	Frequency	Percentage $N = 150$
Low (up to 50 score)	14	9.33
Medium (50–100 score)	117	78.00
High (above 100 score)	19	12.66

Table 10.3 Distribution of farm women according to their extent of participation in agricultural activities

Area of activities	Extent of participation			Mean score
	Low	Medium	High	
Weeding	23	38	89	4.22
Selection of seed variety	35	45	70	4.66
Harvesting	20	38	92	4.16
Soil testing	48	38	64	3.38
Winnowing process	23	30	97	4.43
Seed treatment	34	44	72	3.69
Plant protection	35	34	81	3.92
Sowing	15	41	94	4.40
Manure and fertilizer applications	37	84	29	2.52
Threshing process	29	95	26	2.50
Seed processing	53	39	58	4.84
Grain storage	21	28	101	4.55
Irrigation management	25	89	36	2.79
Collection of harvested crops	31	26	93	4.27
Preparation of land	27	74	49	3.12
Marketing	8	83	59	3.52
Soil treatment	71	37	42	2.64
Overall average	535 (20.98)	863 (33.84)	1152 (45.1.7)	4.64

decision making (20.07%) also found that the more number of farm women in agricultural activities had low involvement in decision making. The activities in which farm women achieved the highest score at low level of decision-making process were weeding (mean score 2.21), selection of seed varieties (mean score 2.46), harvesting (mean score 2.26), soil testing (mean score 2.10), winnowing process (mean score 2.12), seed treatment (mean score 2.42), plant protecting (mean score 2.34), sowing (mean score 2.24), manure & fertilizer application (mean score 2.12), threshing process (mean score 2.28), seed processing (mean score 251), grain storage (mean score 2.31), irrigation management (mean score 2.75) and preparation of land (mean score 2.51), marketing (mean score 1.58), soil treatment (mean score 1.58), and collecting of harvested crops (mean score 1.47).

Hence, there is a need to involve farm women in decision making at higher level and give them opportunity to express their views. The study of Dudi and Meena (2017), Kalpana (2000), Goswami et al. (2004), Praveena (2005), Kavita (2006), Bhattacharjee (2015), Kumar et al. (2016), Pandey et al. (2014) and Tuli et al. (2015) also found that the more number of farm women in agricultural activities has low level of involvement in decision making in this respect in the present study in conformity with these findings (Table 10.5).

Table 10.4 Distribution of farm women according to their decision-making pattern in agricultural activities

	Decision-making pattern			
Area of activities	Low	Medium	High	Mean score
Weeding	92	27	31	2.21
Selection of seed variety	78	35	37	2.46
Harvesting	81	39	30	2.26
Soil testing	92	31	27	2.10
Winnowing process	97	24	29	2.12
Seed treatment	81	33	36	2.46
Plant protection	93	21	36	2.34
Sowing	79	42	29	2.24
Manure and fertilizer applications	85	39	26	2.12
Threshing process	73	48	29	2.28
Seed processing	31	92	27	2.51
Grain storage	77	42	31	2.31
Irrigation management	47	63	40	2.75
Collection of harvested crops	99	31	20	1.47
Preparation of land	39	82	29	2.51
Marketing	89	34	27	1.58
Soil treatment	95	27	28	1.55
Overall average	1328 (52.07)	710 (27.84)	512 (20.07)	2.28

Table 10.5 Constraints faced by farm women

Problems	Frequency	Percentage	Rank $N = 150$
Higher time consumption for household works	65	43.33	V
No management of time for farm and home activities	72	48.00	IV
Family norms	41	27.33	XII
Poor economic status of the family	63	42.00	VI
Male dominance	99	66.00	I
Lack of education	84	56.00	III
Inability to take decision	49	32.66	VIII
Lack of poor guidance for taking decision	42	28.00	XI
Lack of information about the solution to problem	61	40.66	VII
No knowledge about improved technology	85	56.66	II
Higher cost of farm material	46	30.66	IX
Lack of marketing facilities	43	28.66	X
No permission to take decision due to younger age	40	26.66	XIII

10.3.3.1 Constraints Faced by the Farm Women in Decision-Making Process in Agriculture and Suggestions to Overcome Them

The data in tables indicated the constraints faced by farm women in decision making in Agriculture. The problems as reported by farm women are presented on the basis of responses as reported by the respondents. The frequencies of different respondents were worked out on a percentage basis and ranked accordingly. The result indicated that 66.00 percent respondents reported male dominance in decision making in agriculture. This was the major problem faced by them. The second major problems reported by 56.66 percent of respondents were having no knowledge about improved technology. The third problem reported by 56.00 percent respondents was lack of education. The fourth problem reported by 48.00 percent respondents was no management of time for farm and home activities. The fifth problem reported by 43.33 percent respondents was higher time consumption for household works and 42.00 percent of respondents have poor economic status of the family. The seventh problem reported by 40.66 percent respondents was lack of information about the solution to problem, some other problems received by the respondents were inability to take decision, high cost of farm material lack of marketing facilities, lack of proper guidance for taking decision, family norms, and no permission to take decision due to younger age ranked VIII, IX, X, XI, XII, and XIII, respectively.

10.4 Conclusion

The finding of the study shows that majority of the farm women belonged to middle age group, illiterate who were having medium sized family, marginal size of land holding, less social participation, low material resource, low economic motivation, low mass media exposure, whereas innovative proneness, scientific orientation, market orientation, extension contact were found to be of medium level. Further majority of the farm women indicated medium role in decision making followed by high and low level of participation in decision making. It may be also concluded that high number (45.17%) of farm women were found to have overall high level of participation in agricultural operations. On the other hand, the farm women participated in decision-making process in each and every farm activities. The data clearly indicated that the higher number (52.07%) of farm women were observed in low category of decision-making process. As far as constraints faced by farm women in decision making in agriculture concerned, male dominance, no knowledge about improved technology, lack of education, no management of time for farm and home activities, higher time consumption for household works, poor economic status of the family, lack of information about the solution to problem, inability to take decision, high cost of farm material, lack of marketing facilities, lack of proper guidance for taking decision, family norms, and no permission to take decision due to younger age remained the main triggering factors.

References

Badiger PL, Huilgal HP (2004) Participation of farm women in agriculture and animal husbandry. Indian Res J Ext Educ 4(1&2):1124–1128

Begum A (2002) Views on women's subordination and autonomy: Blumerg re-visited. Empowerment 9:85–96

Bhattacharjee D (2015) Participation of women in agricultural activities in Gazipur District of Bangladesh. Indian Res J Extn Edu 15(3):43–46

Chayal K, Dhaka BL (2011) Analysis of role performance of women in farm activities. Indian Res J Ext Edu 10(2)

Doss C et al (2011) The role of women in agriculture, Agricultural Development economics Division, The food and Agriculture organization of United nations. Available at www.FAO.Org/economic/esa

Dudi A, Meena ML (2017) Participation and decision making pattern of farm women in agriculture. Asian J Home Sci 12(1):109–113

FAO (2011) The state of food and agriculture, women in agriculture, closing the gender gap for development

FAO (2003) The uplift and empowerment of rural women in Pakistan. FAO, Rome

Ghosh J (2000) Perception of drudgery by farm women. Empower News 27:1–2

Goswami SN, Challa O, Gawandi RS (2004) Eole of farm women in land management decision making in Vidarbha region of Maharashtra. MANAGE Extn Res Rev 5(1):74–79

Habib N (2000) Invisible farmers: a study on the role of women in agriculture and the impact of pesticides on them in Punjab. Khoj Res. and Public Centre, Lahore, pp 4–5

Hedges TR (1963) Farm management decision. Prentice Hall, Englewood Cliffs

Kalpana M (2000) Role of women in farm and home decision M.Sc. (Ag.) Thesis, Dr. Punjabrao Deshmukh Krishi Vidyapeeth, Akola (M.S.) India

Kavita R (2006) A study on the women participation of farm operations and decision making in agriculture. Ext Educ Mgmt 15(4):26–30

Kumar V, Bairathi R, Sharma AK (2016) Participation and problems of tribal farmers in employment generation activities: an analytical study. Indian Res J Extn Edu 16(1):139–142

Mumtaz KA, Aysha N (1982) Status of rural women in India. Ashish Publishing House, New Delhi, pp 16–34

Pandey S, Kumar N, Singh SK (2014) Participation pattern of rural people in watershed development programme in Haryana State. Indian Res J Extn Edu 14(3):39–42

Praveena V (2005) Decision making pattern of rural women in farm related activities. Agric Extn Rev 3:3–7

Rahman SA (2008) Women's involvement in agriculture in northern and southern Kaduna State. J Gend Stud 17:17–26

Raju VT, Rani S (1991) Decision-making role of women in agriculture. Indian J Home Sci 20 (1):13–17

Sharma S, Badodia S (2016) Impact of participation of rural women in agricultural activities. Indian Res J Extn Edu 16(2):12–14

Singh SP, Gite LP, Kumar N, Agarwal N (2005) Involvement of farm women of Vindhyan Plateau. Agric Ext Rev 1:20–26

Slovenia MM (1998) Women and family in rural development, New Delhi

Tuli AN, Afrad MSI, Haque ME, Bhattacharjee D (2015) Women's participation in household activities in Gazipur district of Bangladesh. Indian Res J Extn Edu 15(3):33–38

Yadav JP, Sharma K, Saini H (2005) Role performance of farm women in animal husbandry practice. Abstract 3rd national extension education congress. April, 27–29, 2005. Organized by Society of Extension Education Agro & NDRI, Karnal, pp 111–112

Chapter 11
Empowering Women Farmers Through Drudgery Reduction and Nutritional Diversity: A KVK, Nalanda-ICAR (Government of India) Initiative

Jyoti Sinha, R. K. Jha, Kumari Vibha Rani, Kumari Punam Pallavi, Seema Kumari, and Anita Kumari

Abstract Agriculture sector employs 85% of rural women in agriculture, yet only about 13% of them possess their own land. There is empirical evidence that women have a decisive role in ensuring food security and preserving local agro-biodiversity. They contribute in every operation of agricultural activities and are also active in allied sectors like cattle management, dairying, beekeeping, goat rearing, mushroom production, poultry farming, etc. Rural women own greater responsibility for the integrated management and use of diverse natural resources to meet the daily household needs. Economic Survey 2017–18 says that with increasing migration of men, there is "feminization" of agriculture sector, number of women farmers are performing multiple roles as cultivators, entrepreneurs, laborers and dominant at all levels of production, pre-harvest, post-harvest, processing, packaging, marketing and agricultural value chain. Agriculture sector in Bihar becomes highly feminized and the situation is more worse that only 7% women are having land rights, though they play a significant and crucial role in agricultural development and allied fields. Livelihood analysis of rural women identifies a "triple burden" of work in the productive, reproductive, and social spheres. This chapter assesses factors influencing the women drudgery in agriculture sector, possible interventions and recommendation for reducing such drudgery either by use of women-friendly tools to make existing tasks easier or increasing the productivity of existing labour or even by changing farm practices with new technology, where KVK plays an important role to impart external inputs and extra knowledge and teaching new technology.

J. Sinha (✉) · K. V. Rani · K. P. Pallavi
Krishi Vigyan Kendra, Nalanda, Bihar, India

R. K. Jha
DRPCAU, Pusa, Bihar, India

S. Kumari
Krishi Vigyan Kendra, Purnia, Bihar, India

A. Kumari
Krishi Vigyan Kendra, Khagaria, Bihar, India

S. Munshi, M. Singh (eds.), *Women Farmers: Unheard Being Heard*, Sustainable Agriculture and Food Security, https://doi.org/10.1007/978-981-19-6978-2_11

Another crucial point assessed by KVK is that there is an orientation gap noticed where women and men face various challenges due to climate change and show different adaptable capacities to manage it. It was noticed that along with productivity nutrition is imperative. Nutrition-sensitive agricultural practices have to be revived. Innovative ideas like Nutrigarden and Mushroom production to promote and establish an enabling environment within health, nutrition, and accessibility of food through agricultural intervention facilitating effective implementation for food and nutritional security through diet diversity are also being promoted by KVKs. Experience and observations in the chapter show that capacity building on digital and financial literacy of women farmer and gender-specific intervention will help in accelerating their confidence and productivity to ensure rural transformation.

Keywords Feminization · Farm · Labourers · Livelihood · Nutrition · KVKs

11.1 Introduction

The Constitution of India gives women equal status to men. If India's dream of inclusive growth has to be realized, women, especially in the country's rural areas, need to be empowered. It means facilitating women with the power of choice in social, economic, political, caste, and gender-based decisions which in turn leads to their own decision making. Given a chance women have the full capacity to make right decisions in relation to their health, education, work, and income generation. This also leads to inclusive economic growth of the country. When it comes to agriculture sector we need to explore in detail. According to Census 2011, women working force in India, 55 percent were agricultural laborers and 24 percent were cultivators. But when it comes to the ownership of the operational property, it is owned by only 12.8%. In spite of women performing numerous labour-intensive jobs such as weeding, hoeing, grass cutting, picking, cotton stick collection, separation of seeds from fiber, livestock production, and other associated activities, still she remains invisible. Her contribution often remains ignored due to patriarchy (Source: TNAU Agritech portal).

11.2 Feminization in Agriculture

In recent years there has been feminization of agriculture in India. The agriculture sector employs 80% of all economically active women in India; they comprise 33% of the agriculture labor force and 48% of the self-employed farmers. In India, 85% of rural women are engaged in agriculture, yet only about 13% own land (Source: https://www.oxfamindia.org/women-empowerment-india-farmers). One of the important reasons behind it remains the larger migration of men through cities, metropolitan for better livelihood generation. This also results in increasing number of women in the value chain.

With growing rural to urban migration by men, there is "feminization" of the agriculture sector resulting into an increasing number of women in multiple roles as cultivators, entrepreneurs, and laborers (Economic survey of India 2017–2018).

Hence, it becomes necessary to experiment innovations which creates opportunity for women's increased access to resources like land, credit, water, seeds, and markets. It is also important to develop need-based technology and strengthening an inclusive extension system focusing on empowering farm women. While working in the KVK system closely with women farmers, we found that women are engaged in diverse farm operations like agriculture farming, animal rearing, poultry rearing, fish farming on a day-to-day basis.

They are involved in a diverse allied farming operations which includes sowing of seeds, transplanting paddy, vegetable cultivation, plucking and selling, weeding, harvesting, winnowing of crops; cutting, collection, carrying and chaffing of fodder to feed the cattle; milking, processing of milk products, cattle shed cleaning, collection of cow-dung, making cow-dung cakes and manures from cow-dung; caring for sick animals, collection of eggs; mushroom production and selling, fish selling, keeping the nets, storage of grains, processing and value addition of seasonal fruits and vegetables at household level, etc. as a day-to-day household operations.

Women Focused Extension System: A Way Forward for Mainstreaming Women in Agriculture! Traditionally in the agriculture system it has been found that the focus has been on training men with knowledge and skills around information and providing access to inputs and services around farming system. Women have remained an ignored entity as far as extension system and trainings are concerned (Staudt 1975; Perraton et al. 1983; Jiggins et al. 1997; FAO). Women-centric agriculture extension efforts allow them to shift their efforts in the right direction and establish their credibility in the rural society. In India understanding the need of an inclusive extension system, Indian Council for Agriculture Research (ICAR) and Krishi Vigyan Kendras (KVKs) have taken lead role for recognizing the importance of it. KVKs are working as center for capacity building, especially for women. It also synthesizes human capacity with technology viability. Advancement of extension techniques for farm women has improved their work in different farming systems and farm enterprises. The same pattern was observed from the paid women and total women (family + paid labour).

Women frequently experience more fragile admittance to beneficial assets, resources, and dynamic force inside the family, lower and less convincing interest in neighborhoods. They have less frequently reach to resources. This can make it hard for women to execute their thoughts and to follow up on the proposals. We must note that rural women produce about 60–80% of the food (Source: Oxfam India).

11.2.1 KVKs Effort Toward Inclusive Agriculture System!

ICAR institutes & KVK have been trying to introduce new techniques for drudgery reduction by providing time and labor-saving women-friendly tools. Women needs and aspirations are at the center of such interventions. There are trainings designed oriented toward skill development and capacity building enduring higher productivity, nutrition, health security, food security, economic security, and livelihood security of farm women.

It is closely experienced and observed that if the farm women are given handholding supports, trainings with improved scientific knowledge, better management practices, and on time information on agro-advisories it helps them to be more productive.

Women Farmers and Drudgery Women do not work in a standalone fashion in their household. It is important to understand that along with the work at farm the women have major chunk of time spent on household chores (Lanjouw and Lanjouw 2001).

Understanding the need KVKs have been proactive in deciphering intervention and introducing training/skill building programs which can help the women farmers to be more productive in less time and help the reduction in drudgery. Introducing women-friendly and simple tools, like double wheel weeder for simple agricultural operation, helped in time saving and drudgery reduction.

Study 1: Performance of Traditional Method and Hand-Operated Groundnut Decorticator Nalanda, Bihar Data collected in a field demonstration of Nalanda district of Bihar, four groups comprise 40 women between the ages of 30 and 50 years participated in Frontline Demonstration on manually operated groundnut decorticator (CIAE, Bhopal model) conducted by Krishi Vigyan Kendra, Nalanda. Adoption of this technology showed an increase in productivity in terms of money by selling deshelled kernels. The gross cost calculated for a day by the traditional hand shelling method was Rs. 624.00, including labour charges for winnowing, cleaning, and storage material and gross return was observed as Rs. 654.00 for total 5.45 finished kernels @ Rs 120/kg. Whereas the decorticator produced 32.4 kg/h with gross costing Rs.20,100/- for semiskilled labour, storage materials, raw materials with the machine working for 6 h/day. The gross return calculated Rs.23,200 for a total 193 kg @Rs.120 added with cost of broken kernels@Rs.60/kg. Hence, BC ratio calculated was 1:1.15. Groundnut deshelling is tedious and time-consuming, with low performance of shelled peanuts approximate 1.2–1.4 kg/h. In comparison, hand-operated groundnut decorticator showed 32.4 kg/h with mean efficiency 66.4%. Ogunwole (2013) mentioned the manually operated Sheller with roasted groundnut sample performed at 55% shelling efficiency. If we go on continuous work on machine, we got greater output in very short time. If we put input in large amount, then we got the more output from machine with a minimum wastage (Raghtate and Handa 2014).

	No. of demo	Weight of g'nut seeds (in kg)	Mean Wt of peeled g'nut seeds (kg)	Mean Wt of broken g'nut seeds (kg)	Effective time of peeling (min)	Peeling efficiency (%)	Av. Production in kg/h
Deshelling by hand	20	2.0	1.24	0.08	51.65	62.17	1.44
Hand-operated groundnut decorticator	20	2.0	1.33	0.15	2.46	66.42	32.43

Observation for using simple machine regarding saving of time with less tiresome practices could be utilized by women farmers for other additional farm activities, taking care of children and households or investing to self-health as benefits as taking rest. The productivity gain and friendly tool under demonstrations practices over conventional practices created awareness and motivated other women farmers to adopt new implement with the result of value-added products of groundnuts in the district.

Study 2: Drudgery Reduction by Introduction of Farmer Friendly Tools Another study done as farm trial on how drudgery responsible for women's exploitation and impacts on physical, mental, and psychological strata at KVK, Manjhi, Saran on methods and tools for reducing drudgery in women, by introducing women-friendly tools wheeled weeder and long stick hand hoe for weeding operation showed result for area covered 26.89 (m^2)/h/labor in farmers practice, 67.44(m^2) by long stick hand hoe and 56.94 (m^2) by Wheeled weeder. The BC ratio was found 1.50 and 1.118, respectively, resulting that instrument reduces the cost involved in weeding operation apart from reduction in physical stress of the women labour.

(a) Work efficiency estimation through visual observation and qualitative perception

Observations	Farmers practice (Weeding by Khurpi)	Weeding by long stick hand hoe	Weeding by wheeled weeder
Cleanliness of the field	Excellent	Good	Satisfactory
Mulching effect	Excellent	Satisfactory	Good
Earthing up	Good	Satisfactory	Excellent
Area covered	Satisfactory	Good	Excellent
Cost involved	Satisfactory	Good	Excellent
Rouging	Excellent	Good	Satisfactory
Total score	13	10	13

Satisfactory 1, Good 2, Excellent 3

(b) Study of physical strain by questionnaire/schedule:

Technology assessed/ refined	Area covered/hour/labour (m^2)	Cost involved in Rs./ha at 20 DAS	BC ratio
Farmer's practice	26.89	3162.90	0
Long stick hand hoe	67.44	1260.40	1.509
Wheeled weeder	56.94	1493.03	1.118

Treatments	Observations									
	Feeling physical strain after 1 h of working in stretch	Feeling physical strain after 2 h of working in stretch	Feeling physical strain after 3 h of working in stretch	Feeling physical strain after 4 h of working in stretch	Need to change the position between 1 to 2 h of working	Need to take break or rest at 2 h	Feeling stress on fingers	Feeling stress on waist portion	Feeling stress on shoulders	Feeling stress on palm
Farmers practice (weeding by Khurpi)	0.099a	0.297a	0.594b	0.693b	0.495a	0.693a	0.693b	.693a	0.693b	0.693b
Weeding by long stick hand hoe	0.000	0.099a	0.297a	0.495a	0.396a	0.594a	0.495a	.495a	0.297a	0.396a
Weeding by wheeled weeder	0.000	0.000	0.198a	0.297a	0.297a	0.495a	0.396a	0.396a	0.198a	0.297a
CD at 5%	0.176	0.269	0.337	0.287	0.359	0.238	0.269	0.329	0.380	0.387
SEm(±)	0.057	0.087	0.109	0.093	0.116	0.077	0.087	0.106	0.123	0.125

11.3 Women Farmers and Nutrition

In the work carried closely with the women farmers it was found that along with productivity nutrition is imperative (https://blogs.worldbank.org/health/nutrition-and-agriculture-bridging-gap).

In recent years progress has been made in developing knowledge and understanding the magnitude and scope of nutritional challenges. The existing problem of malnutrition has taken major shift to identify micronutrient deficiencies thereby certain diet modification needs to incorporated by establishing nutrition gardens as health and nutrition security approaches in KVKs Nutrition-sensitive agricultural practices has to be revived. KVKs are focusing on the reintroduction of nutri-cereals, nutri-garden, backyard poultry, mushroom production, etc. to enrich their diet. Many efforts have been made to incorporate local food into regular dietary practices to bring out positive behavioral changes in people's mindsets. There are needs for innovative ideas to promote and establish an enabling environment within health, nutrition, and accessibility of food through agricultural intervention facilitating effective implementation for food and nutritional security through diet diversity. Several supply-side food and agricultural interventions suggest promise in improving nutrition. Interventions that target micronutrient deficiency are often considered to be the most cost-effective public health programs. With experience we have found that introducing mushroom production and nutri-garden with each rural household is the most promising concept for crop diversification. The positive aspect is its availability at the doorstep of the farmers. It has also been regarded as an effective means for poverty alleviation due to its potential for a quick, high return of profit on a small investment, increasing the purchasing power of rural households (https://zaszambia.files.wordpress.com/).

Mushroom being high nutritive value crop contains adequate protein and many micronutrients. It is quickly produced on locally available materials, without land and minimum input and can have accessibility to the most disadvantaged population.

Indeed, the evidence suggests that nutrition-sensitive food and agricultural interventions can increase nutritious foods and improve dietary diversity. Thus, identifying and making them understand the ignorance and nutritional gap of high-quality low-cost food accessibility is imperative. Changes in dietary habits and low cost intervention at home level, self-produced, and nutritious food crop as oyster mushroom can fulfill the nutritional need. In one of the projects implemented by KVK by KVK Nalanda and supported by Integrated Child Development Scheme, Govt. of Bihar on mushroom production and behavioral changes for better nutrition has strongly generated evidences where mushroom production done by women farmers had positive effect on the nutritional consumption of food at household level and also created a level playing field for them for added income generation.

The association of KVKs and ICDS had a multipronged effect increasing the nutrition for children and household, creating an entrepreneurial scope for women farmers, additional income generation, and improved knowledge as a farmer.

The intervention created a considerable impact of mushroom production with Anganwadi participation resulted in a multidimensional effect.

1. Household mushroom production enterprise secured fresh and easy to reach good quality food.
2. Inclusion of mushroom in their regular diet resulted inadequate protein and micronutrients supplementation regularly even in lean periods like in COVID-19 pandemic period.
3. Increasing demand for mushroom led rural women and youth to set in-house production units for income generation activity and maintain local food availability.
4. Processing and dehydration of mushroom could enable them to reduce wastage, whereas value addition in different local recipes provided variety and satiety to food.
5. Increased frequency noted in mushroom consumption and related nutrient intake, which will significantly result in a better growth rate in children and pregnant ladies.

Impact	Before	After introducing mushroom production
Consumption in no. of families	35	290
Frequency/week	1–2	4–5
Quantity in g/head	75–100	150–180
Inclusion in diet through different recipes	02	09
Monetary gain/Entrepreneurship development in villages (including other than surveyed community)	04	34
Earnings (profit)	–	Rs30/kg
No. of unit established	25	53
No. of queries/training demand coming up	13	40

Same way nutri-garden units developed at selected Anganwadi centers under umbrella project of Bihar Agricultural University Sabour, "Apni Kyari Apni Thali" has showed that consumption of green leafy vegetables and fruits increased in the diets of pregnant, lactating women and children of 6–24-month-old for supplementary feeding. Supply of selected nutritious crops, garden tools, method of cultivation, uses, and preservation techniques were facilitated by KVK, Khagaria to ensure their availability throughout the year. This program was aimed to reduce at least Rs.5000/- for fruits and vegetables from the total food budget of AW center. The Nutri-garden at Anganwadi Centers utilizes locally available compost for promotion of zero budget vegetable cultivation. This area's local women have been associated as committee members to the nutri-garden for production, management, and efficient utilization.

Types of vegetables/ fruits	Amounts of vegetables/fruits consumption (in grams) before and after implementation of the project								
	Pregnant women			Lactating women			Children's of 3–6 years		
	Before	After	Standard (ICAR)	Before	After	Standard (ICAR)	Before	After	Standard (ICAR)
Green leafy vegetables	110	140	150	120	135	150	20	23	30
Root vegetables, etc.	69	71	75	62	69	75	19	24	30
Other vegetables	52	65	75	55	67	75	15	19	30
Fruits	16	18	30	13	19	30	25	33	50

[a] The presented data is pooled from KVK, Khagaria and KVK, Purnea

Experience and Observations of Working with Women Farming Community in Rural Bihar There is a gender equity gap where women and men have different vulnerabilities to climate change and different adaptive capacities to deal with it (Ashby et al. 2012).

As a scientist working in KVKs it is important that we have purposive sampling of women farmers in our interventions starting from capacity building programs, skill development, technology intervention, on farm trials or market linkage. If women had the same access to productive resources as men, they could increase yields on their farms by 20–30 percent (FAO2011). Hence, it becomes increasingly important to involve women as far as possible through innovation at the grassroots level by incorporating it in KVKs activities. We have seen in allied agricultural activities like mushroom production, vermicomposting, goatery and backyard poultry, women participation go upto 60–80%.

As per data observed from farmers registration done for soil sample analysis at KVK, nearly 5% of women registered their name for soil testing). Women who received trainings in processing and value addition of fruits, vegetables, and mushroom, were only indulged at home level production which definitely led to home level food security but not as a sustainable income generation activity. In spite of that there is an abundance of raw material and Govt. schemes available, only 1–2% group started commercial ventures. Packaging and marketing of fresh and processed food items or other agri-produce like bio- pesticides and vermicompost etc always remain a challenge. Association with the SHGs, makes women more confident and they are able to understand bank operations to avail credit facilities (Kaur et al. 2017). But still they have inhibitions in starting independent business. In today's digital world, women seem to be far behind and thats an additional hinderance in farming. There is an increasing need for initiative which should focus on their financial and digital literacy developing their skills related to the use of simple technical apps and software. It is reported by NFHS 2015 survey that nearly 96.1% of the women access to mobile phones in urban and 87.3% in rural area but still only 20.4% could use internet in urban and 5.7% in rural India (Smitha 2021).

There has to be an understanding that financial literacy of farmer women will help in accelerating their confidence and productivity. Though the literacy rate among urban women is 79.01 percent and among rural women is 57.39 percent still India stood at the 23rd place in respect of financial literacy, having only 35 percent of Indians as financially literate (Visa, Financial Literacy Survey, 2014). According to S&P survey, 2015 only 20 percent of Indian women are financially literate to take benefits from it. Illiteracy, lack of infrastructure, and dearth of awareness are the roadblocks in the path of financial inclusion (Sibi and Ananth 2017; Arora et al. 2018).

References

Sibi MS, Ananth AA (2017) Banking inclusion—a gateway to financial inclusion. Sumedha J Manag 6(1):4–19. Visa, Financial Literacy Survey, Indians among least financially literate people globally, Firstpost.com, December 2014

Arora B, Sharma S, Pahwa MS (2018) Empowering women through financial literacy in Rajasthan. Int J Computer Sci Eng 6(9) E-ISSN: 2347-2693

Ashby J, Kristjanson P, Thornton P, Campbell B, Vermeulen S, Wollenberg E (2012) CCAFS gender strategy. CGIAR Research Program on Climate Change, Agriculture and Food Security (CCAFS), Copenhagen. Available online at: www.ccafs.cgiar.org. CR FARNWORTH &K ECOLVERSON J Gender. Agric Food Security 1(1):20–39, 2015

Kaur L, Sachan D, Sulibhavimath A (2017, April 07). Women empowerment through self help groups: case study in Jalandhar District of Punjab IJBSM [Internet]. [cited 8Feb.2022];8 (1):340–345. Available from: http://www.pphouse.org/ijbsm-article-details.php?article=942

Smitha HS (2021) Empowerment of women through digital literacy strategies—new challenges. IOSR J Humanit Soc Sci. e-ISSN: 2279–0837, p-ISSN: 2279-0845. www.iosrjournals.org:21–23

Ogunwole OA (2013) Design, fabrication & testing of manually & electrically operated roasted groundnut decorticating machine. Food Sci Qual Manag 14:1–10. www.iiste.org ISSN 2224-6088 (Paper) ISSN 2225-0557 (Online). Published Online March 2013 (http://www.scirp.org/journal/ojop)

Raghtate AS, Handa CC (2014) Design and fabrication of groundnut sheller machine. Int J Innov Res Sci Technol 1(7) ISSN (online): 2349-601

India National Family Health Survey, NFHS-4 (2015–16) [FR339], Table-2.8, pg 34

https://www.oxfamindia.org/women-empowerment-india-farmers (2018) Move over 'Sons of the soil': why you need to know the female farmers that are revolutionizing agriculture in India, Gender Justice, 15 Nov 2018

Jiggins J, Samanta RK, Olawoye JE (1997) Improving women farmers' access to extension services, *Improving Agricultural Extension. A reference manual*, FAO 1997

Perraton H, Jamison D, Orival F (1983) Mass media for agricultural extension in Malawi. In: Basic education and agricultural extension: costs, effects and alternatives. Staff Working Paper No. 564. World Bank, Washington, pp 147–202

Staudt K (1975) Women farmers and inequities in agricultural services. Rural Afr 29:81–93

Lanjouw JO, Lanjouw P (2001) The rural non-farm sector: issues and evidence from developing countries. World Dev 26:1–23

Chapter 12
Women Farmers and Technologies in Agriculture: A Review of Current Practices

Nidhi Thakur

Abstract Women play an important role in the agriculture sector and development of rural economies. Their work is diversified as farmers, laborers, and entrepreneurs. Gender mainstreaming and empowerment of the women farmers is a key concern for which various policies and solutions are being explored. It is widely observed that rural women are getting more and more marginalized due to several reasons such as less investment in agriculture, lack of decision-making authority, inability in accessing the economic opportunities and technologies. There are several systemic challenges that lead to gender disparities in agriculture; however, the role of technologies is well recognized in overcoming the constraints of low productivity, access to information, and drudgery. Most of the agri-technologies can create new employment opportunities and better access to organized markets and cooperatives for rural women and can significantly increase the efficiency and effectiveness in rural women enterprises. However, there are large gender disparities in the adoption of such technologies. Constraints such as sociocultural norms about gender roles, lack of information, prevailing digital-divide, limitations in decision-making power, lack of finances, lack of agency for hand-holding, and lack of resources to implement policies are prominently noted in several studies. Technologies supported by appropriate institutional arrangements and models of linkages can empower women farmers to address major socio-economic, gender, and environmental issues. Therefore, it is imperative to design strategies that promote gender-equitable outcomes and can be used to mainstream gender in agricultural technology adoption.

Keywords Rural · Women · Agri-technologies · Decision-making · Policy · ICT

Supplementary Information The online version contains supplementary material available at https://doi.org/10.1007/978-981-19-6978-2_12.

N. Thakur (✉)
Office of the Principal Scientific Advisor, Govt of India, New Delhi, India

S. Munshi, M. Singh (eds.), *Women Farmers: Unheard Being Heard*, Sustainable Agriculture and Food Security, https://doi.org/10.1007/978-981-19-6978-2_12

12.1 Background

Women constitute a large part of the agriculture workforce. The agriculture sector employs 80% of all economically active women in India; they comprise 33% of the agriculture labor force and 48% of the self-employed farmers. In India, 85% of rural women are engaged in agriculture, yet only about 13% own land (Oxfam 2018). As farmers, laborers, and entrepreneurs, women play an important role in the agriculture sector and the development of rural economies. Economic Survey (2017–2018) points to the "feminization" of the agriculture sector, with an increasing number of women in multiple roles as cultivators, entrepreneurs, due to the migration of men to the urban areas. Table 12.1 refers to the gender wise distribution of farmers across different regions of the globe and Table 12.2 depicts that agriculture sector has the highest share workers and the number of women workers/farmers has remained higher than men over the period of time. Despite such strong representation in the workforce, gender disparity is deeply rooted in the agriculture sector, as is the case with most sectors of the economy. Technology can be crucial in supporting women's participation in higher-value activities in global agri-value chains. Several studies (Murray et al. 2016; Sinyolo 2020) have shown that increasing technology adoption amongst women farmers could result in increased economic benefits at the grass-roots as well as significant social impact. However, women face several gender-

Table 12.1 Gender-wise percentage distribution of workers across sectors in rural areas

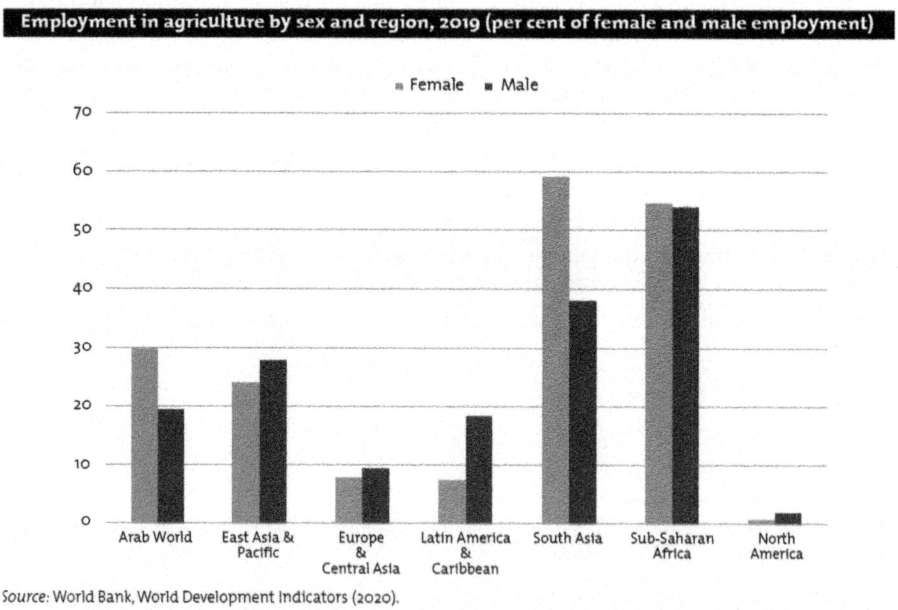

Table 12.2 Percentages between 2001 and 2011 of cultivators and agriculture labourers (men and women)

	1999–2000[a]		2004–2005[a]		2011–2012[a]		2017–2018[b]	
Sectors	Male (%)	Female (%)	Male (%)	Female (%)	Male (%)	Female (%)	Male (%)	Female (%)
Agriculture	71.4	85.4	66.5	83.3	59.4	74.9	55	73.2
Mining and quarrying	0.6	0.3	0.6	0.3	0.5	0.3	0.5	0.2
Manufacturing	7.3	7.6	7.9	8.4	8.1	9.8	7.7	8.1
Construction	4.5	1.1	6.8	1.5	13	6.6	14.5	5.3
Trade, hotel, and restaurant	6.8	2	8.3	2.5	8	3	9.2	4
Other services	9.4	3.6	9.9	4	11	5.4	13.1	9.2

Source: [a]NSS rounds; [b]MoSPI (2019)

based barriers to accessing technologies, and in most cases the available technologies often fail to meet their needs.

In many parts of the world, it is common for men and women to take on different roles in agriculture production, processing, and marketing and consequently use different types of technology or use the same technologies to different degrees. Women perform a variety of agricultural operations physically, mostly drudgery-filled operations such as sowing, transplanting, weeding, scaring birds, harvesting, threshing, winnowing, drying, and storing of seeds and food grains. The use of several context-specific labor-saving technologies in agriculture can be instrumental in reducing women's time and work burden in agriculture (Van Eerdewijk and Danielsen 2015; Vemireddy and Pingali 2021). In most places, the entire livestock production activities are managed by the womenfolk. Tending animals, collecting, carrying, and cutting fodder and feed for animals, milking the animals and marketing, preparing milk products, cleaning the cattle shed as well as cattle are attended to exclusively by the rural women. It has been noted in several research studies that women lag behind men in agricultural technology adoption (Rola-Rubzen et al. 2020) due to barriers like access to capital, access to inputs and services (information, extension, credit, fertilizer), physical accessibility, and cultural norms (Achandi et al. 2018). The Committee constituted by the National Commission for Women in early 2000 (Committee Report 1998) observed that "Technology Empowerment of Women in Agriculture" is possible by factoring in multiple roles of women, which necessitates the integration of the gender perspective while considering technological inputs. The provision of support services such as childcare, basic facilities of drinking water, sanitation, fuel, etc. cannot be overlooked while attempting to empower women technologically.

The inequitable access to credit and land resources has put women at a greater risk of poverty and lower productivity within agriculture. Access to agri-technologies will not only reduce their burden and increase their productivity but will also have a positive impact on women and children's health and nutrition. Minimizing the gender gap among smallholder farmers for technology adoption is challenging as most of the technologies do not address gender concerns. Several gender-related nuances of adoption behavior need to be understood, for increasing the technology penetration and consequently raising agricultural productivity and women's welfare.

This chapter discusses agri-tech and the gender biases that may prevent women from accessing economic opportunities through improved technologies and mechanisms, with specific focus on labor-saving technologies (Sect. 12.2), Information and communication technologies (Sect. 12.3), and Sustainable agriculture technologies (Sect. 12.4). A snapshot of Government initiatives and policy measures is also discussed in each of the subsections.

Labor-Saving Technologies The Labor-saving technologies can be defined as "practices or techniques, tools or equipment, know-how, and skills, or combinations of the afore mentioned elements that are used to enhance productivity, reduce production and processing costs and save on scarce resources or inputs, such as labor or energy" (Ragasa 2012). Labor-saving technologies can save efforts in several farm-related activities like Mechanization (shellers, drum seeders, weeders, jab planters, food processors), Inputs for crop farming (drought-resistant seeds; trees for woodlots), Infrastructure (milk coolers, rainwater harvesting reservoirs; biogas units; improved stoves; livestock pens), Transport (animal or energy-based: bicycles or carts), and many others. These technologies and tools could help improve not only the livelihoods of women smallholder farmers but are also important for making rural areas more attractive locations to reside in, thereby reducing the pressure to migrate. There are also clear benefits to the environment and climate change mitigation efforts. Several technologies for reducing Drudgery for women farmers available in India are Hand Ridger, Seed Treatment Drum, Fertilizer Broadcaster, Seed Drill, Rotary Dibbler, Four-Row Paddy Drum Seeder, Tow-Row Rice Transplanter, Twin Wheel Hoe, Groundnut Decorticator, Tubular Maize Sheller, Cono Weeder, Cotton Stalk Puller (Jaw Type), Sugarcane Stripper, Pedal Operated Paddy Thresher, etc. A compendium of such technolgies is published by ICAR- Drudgery reducing technologies for women farmers (ICAR n.d.).

However, the introduction of any agricultural technology and its subsequent adoption by a farming population would normally depend to a large extent on the society's socio-cultural and economic ideologies as well as the application of these technologies to local production systems (IFAD-FAO 2005). There are several other determinants of the adoption of these technologies:

The perceived effectiveness of the technology: Perceived benefits of technology depend, amongst others, on the information available to each decision-maker. Labor-saving technologies in poor rural development settings need to be reliable and bring measurable advantages. Many times marginal improvement than the traditional practices leads to rejection. Further, nonalignment of "performance indicators" between the tech developers and the user base also leads to poor response. In such cases, a dialogue is required between both parties, and behavior change communication can be used to achieve beneficial changes in attitudes and behaviors.

The appropriateness of the technology in daily life: Labor-saving technologies developed or adapted by local artisans and manufacturers in consultation with women users are much more likely to be appropriate in women's lives than those developed in isolation from rural communities (Lambrou and Piana 2006). Women's indigenous knowledge and experience in performing tasks and managing local natural resources needs to be used in designing and development of technologies. As an instance, the farm mechanization technology and tools are designed considering male ergonomics. Gendered considerations, like giving more attention to women-oriented activities (such as weeding and transplanting), the weight of the equipment, ease of use by women, and labor-saving benefits for women, would be useful. The use of local materials and labor to produce technologies supports the local rural economy. Relevant governmental ministries—such as agriculture, rural development, or water—can assist the process by supporting rural artisans and manufacturers who produce technologies and corresponding tools, equipment, and spare parts (Gender in Agriculture Source Book (2009)).

A*lignment with internal decision-making of the rural household:* The household budget holders, which are often men, need to be convinced of the benefits of labor-saving technologies, for successful adoption. It is also important to assess how women's and men's roles and workloads may shift as a result of labor-saving technologies and how this change is perceived. Raising awareness and communication can be very useful to bring about beneficial behavior outcomes.

Awareness and training for adoption: Several studies suggest that improved technologies reduce the drudgery of women, improve the output and comfort. For efficient use, accessibility and training are important. This was demonstrated by introducing a set of 14 drudgery reducing farm tools and implements, i.e. sapling transplanter, sickle/khurpi, long handle weeders, three types of harvest bags, ring cutter, finger guards, milking stand cum stool, head load manager, seed cum fertilizer bag, seed placement tube, and fertilizer broad caster. A capacity-building training program was conducted, in the village of Ramachandraguda, Rangareddy district, Telangana for exposure about the improved set of farm tools and implements. It was demonstrated that the adoption levels increased significantly after the training programs. This study reemphasized the need for educating the farmers about the importance of drudgery reducing technologies and the training sessions. However, the study also indicated that accessibility of tools is very important as farmers specially women may not be able to afford them. It was therefore recommended that the village community center can provide them on a rental basis for the benefit of women farmers (Vijaya and Deepika 2020). Box 12.1 illustrates the importance of awareness and training in deciding the appropriateness of technology for adoption by women farmers.

Labor-saving technologies are particularly attractive when they add value to performing the domestic task in terms of saving time, economic value of the end product, and health outcomes. In some cases, increasing demand for modern technologies, including those that are cleaner and fuel-efficient, may create new income-generating activities for women or their households. Communicating the benefits of adoption to the farmers is very much desired for larger impact.

Government Initiatives in Improving Farm Mechanization and Reduction of Drudgery The government is making several interventions in this regard. During the period from 2014–2015 to 2020–2021, an amount of Rs.3606.72 crores has been allocated for agricultural mechanization. Out of these, 11,62,437 numbers of machines and equipment have been provided to farmers on subsidy, 10,209 custom hiring centers, 255 high-tech hubs, and 7828 farm machinery banks have been established. A central sector scheme was launched in 2018 for Punjab, Uttar Pradesh, Haryana, and Delhi to manage crop residue. During the year 2019–2020, Punjab, Haryana, Uttar Pradesh, and National Capital Territory of Delhi have provided assistance to individual farmers for distribution of 12,717 in-situ crop residue management machinery and 8866 custom hiring centers. "Farms" (Farm Machinery Solutions) App (CHC—Farm Machinery Mobile App Advanced Edition), a multilingual Android platform, were launched in order to bring together all the agricultural machinery custom service providers and farmers/users, on a single platform (PIB Press release). The Sub-Mission on Agricultural Mechanization (SMAM) has a suitable platform which consolidates all activities related to agricultural mechanization and provides a "single window" approach for implementation with accelerated and inclusive growth of agricultural mechanization in India. The scheme is implemented in all the states, to promote the usage of farm mechanization (https://agricoop.gov.in/sites/default/files/SMAM%20%282%29_1.pdf).

However, there are several fundamental concerns to availing the benefits of these Govt. programs & schemes such as access to finance, maintenance of machinery, technical know-how, and others. These challenges become a major deterrent in the case of women farmers. Mechanization of Indian farms and reduction in drudgery-based operations for women farmers would require several steps. One of these can be to develop a provision for hiring the machines as well as for the mechanized services. This step will reduce the need for each farmer to own the machinery and learn skills to operate the individual machines. Such mechanized service enterprises for farmers will also create jobs for skilled youth in that region.

Traditionally, agricultural technologies had been large and expensive and were affordable to medium and large farmers. Smallholder farmers, particularly women have rarely been the focused audience of technology development. The lack of attention to women's need in technology design and access reduces the possibility of enhanced farm productivity, especially for crops in which women play a predominant role (Hansda 2017). The sustained use of agricultural technologies at a grassroots level requires significant capacity-building exercises and on-site training. Women-friendly labor-saving technologies and improved farm tools need to be popularized among women farmers for their awareness and utilization. The agriculture research centers and extension services should adopt a community-based participatory approach for developing and validating women-friendly technologies, this would also result in getting crucial feedback for further modifications. The capacity-building exercises and awareness programs should have women representation. This will directly impact the participation of women farmers which in turn will affect adoption.

12.2 Digital and ICT Technologies

12.2.1 Information and Communication Technologies Include Devices, Networks, Services, and Applications

These can range from cutting-edge Internet-based technologies and sensing tools to other technologies that have been around for much longer, such as radio, telephones, mobile phones, television, and satellites. ICTs and other digital technologies can improve the lives of small farmers in a myriad of ways, from monitoring crops to tracking market prices and from spreading good practices to facilitating access to banking services. These technologies provide better access to knowledge and information, which will, in turn, facilitate their potential integration into global value chains. Particularly for gender agricultural development, ICTs can empower women with better access to knowledge and information for improved decision making and help women farmers to adopt other new technologies in agriculture as they can be powerful tools for spreading awareness. Adoption of these technologies will create new avenues of employment and promote entrepreneurship (Mathur 2017). However, the access and use of these technologies by women is limited due to several factors: Most of the applications used for information dissemination are specifically targeted to male farmers as they are decision-makers on inputs, choice of crop, farm operations, marketing, availing government schemes and others (Gender Mainstreaming in ICT for Agriculture (2012)). Further, the generic information often lacks contextualization and ignores the special needs of women, for using the information. Other socio-cultural factors that govern access and use of ICTs are asset ownership and gender digital divide, whereas compared to men, rural women are less likely to own digital and communication equipment (Kamala et al. 2019). Due to multiple roles and heavy domestic responsibilities, there is a time constraint to learn ICTs or even access the information. The cultural barriers of limited interaction with extension workers or visiting the community centers add to restricted use of technologies. Meena and Seetharaman (2004) reported the major constraints experienced by farm women in utilizing the media (Radio and Television) program were lack of adequate time (74.17%), affecting studies of children (30%), lack of adequate awareness about media program and less understandability of subject matter (25.83%), not able to take notes (25.80%) in case of farm broadcast alone. Further in case of television affecting studies of children (64.70%), lack of location specific, need based and interesting farm program (61.77%), lack of adequate time (58.33%), and unsuitability of time of farm telecast (31.67%) (Tables 12.1 and 12.2).

12.2.2 Factors for Promoting the Adoption of ICTs and Other Digital Technologies

Digital infrastructure and digital literacy: Though ICTs offer considerable potential for reaching the unreached with relevant information and services, women's ownership and access to ICTs is low. Access to the internet and telecommunications are mostly limited to urban areas in many developing countries while the rural areas remain beyond the ambit of new technology (Kale et al. 2015). Inadequate, and unstable power supply, cost of hardware and software are high with respect to average rural dwellers (Kale et al. 2015). Digital literacy in rural communities should be enhanced by taking into consideration local needs and constraints. The lack of basic skills of using ICT facilities in agriculture causing the inability to deliver adequate ICT knowledge to farmers has been reported (Singh et al. 2018). It's also important to provide a safe environment for learning and sharing of knowledge. Appropriate learning opportunities will enhance individual and collective decision-making skills. Box 12.2 presents an example of how women from a remote village of Bihar were empowered due to digital technologies and platforms.

Ensuring agricultural service delivery systems to marginalized women farmers: Marginal women farmers, particularly those engaged in farm labor on other people's land and forest-produce gatherers, face significant barriers to formal agricultural service delivery mechanisms and informational channels provided by departments and agricultural universities. In a report, on challenges faced by Women Farmers for accessing information, several barriers to information access were recognized such as the loss of a day's work when accessing agencies at the block level, social constraints on movement, and intimidation while completing entitlement applications through official channels (https://itforchange.net/sites/default/files/IT%20for%20Change-InputstoMAKAAMNov2015.pdf). In the era of enhanced connectivity and digitization, mobile phones could hold the potential of removing some of these limitations, which would have been the thought process behind launching several mobile apps. Mobile info-services can bring gains for marginal women farmers only if they are tied to a process where access to schemes and services and other expert institutions and agricultural advisers is facilitated through an on-ground support structure.

Counseling and vocational training: Most rural women lack both the means and the opportunity to obtain a formal education because of poverty. Some women also opt out of training programs because of cultural, religious, or family pressures. To utilize the potential of ICTs, women will need access to credit and training on new and appropriate technologies. ICTs can be useful for skill development and knowledge dissemination in areas where women participation is high such as dairy, sericulture, bee keeping, livestock management, integrating farming systems, and others. Identification of the right blend of technologies which are locally relevant should be adopted to increase the efficiency of initiatives for ICT in agriculture and better serve different users and contexts. The training manuals and programs should be customized as per local needs, e.g. the use of regional language, a mix of graphics and pictures for explaining the concept, community theaters and street plays can be useful. ICT can empower women farmers as individuals, and it will improve their chances of networking in agri-food business like vegetable cultivation, door mat making, fish cultivation, agro processing, etc.

Gender-sensitive approach: it is known that the socio-economic conditions, cultural norms, and poor infrastructure are strong deterrents for rural women to access ICTs. Interventions that are mindful of these inequalities are more likely to be successful, e.g. the transmission of farm advisory that provides information at a time when the women folk are free from household chores and in local language would be much more meaningful. Similarly, lower literacy rates amongst women farmers should be considered when designing a technological intervention. By integrating video, audio, and imagery, instead of text would ensure that women adopt and use these solutions.

Partnerships and local influencers: Strategies that leverage local leaders (women) or early adopters will provide more credibility to the technology. In addition, to increase awareness and outreach of technologies, local influencers can be extremely useful for technology that is seeking to enter new markets. Local organizations, communities, and NGOs often have the social capital to provide trusted information and good quality services. Such partnerships provide a robust platform for adoption and use of technologies which can meet various needs of farmers.

12.2.3 Government Initiatives in Digitalization of Agriculture

Government has taken various initiatives to enable digitalization of agricultural sector in the country and to promote agri-tech business. Some of the initiatives in this direction are mentioned below (https://pib.gov.in/PressReleaseIframePage.aspx?PRID=1777684)

- The India Digital Ecosystem of Agriculture (IDEA) framework which intends to build the federated farmers' database using available data and by linking them with the digitized land records.
- The National e-Governance Plan in Agriculture (NeGP-A) wherein funds are released to the State(s)/UT(s) for project involving use of modern technologies, viz. Artificial Intelligence (AI), Machine Learning (ML), Robotics, Drones, Data Analytics, Block Chain, etc.
- The National Agriculture Market (e-NAM) Scheme with the objective of creating online transparent competitive bidding system to facilitate farmers with remunerative prices for their produce.
- The Centralized Farm Machinery Performance Testing Portal and FARMS Mobile App.
- A Central Sector Scheme of financing facility under "Agriculture Infrastructure Fund" for investment in digital Connectivity and optic fiber infrastructure.
- The Krishi Yantra App, developed by ICAR to augment research, operations and technology dissemination process in the field of agricultural engineering.

There are special provisions in national schemes, especially for the women farmers, like-National Mission on Agricultural Extension & Technology (NMAET), Agriculture Technology Management Agency (ATMA), Agri-Clinics & Agri-Business Centers (ACABC), Mass Media Support to Agricultural Extension, Mission for Integrated Development of Horticulture (MIDH), National Mission on Oilseeds & Oil Palm (NMOOP), Integrated Scheme for Agricultural Marketing (ISAM), National Food Security Mission (NFSM), National Mission for Sustainable Agriculture (NMSA), Sub-Mission on Agricultural Mechanization (SMAM) (https://vikaspedia.in/agriculture/women-and-agriculture/special-provisions-for-women-farmers-in-national-schemes#:~:text=Specific%20provisions%20(only%20for%20Women)&text=0.10%20lakh%20per%20group%2Fyear,at%20least%202%20FSGs%2FBlock).

Though these schemes look very promising and ensure inclusivity, there has been a shift away from targeted agri-extension efforts. ATMA, the pilot testing of new institutional arrangements for technology dissemination at district level and below, intends to be a bottom-up approach, is to make the technology dissemination farmer driven and farmer accountable (https://agritech.tnau.ac.in/atma/atma_intro.html#beneficiaries). This addresses gender concerns to some extent but it is in fact a move toward generic extension program. The e-agricultural initiatives as well tend to ignore the importance of gender in their design. The women-only schemes and programs are scarce.

Digital technologies or ICTs (Information and Communications Technologies) hold tremendous potential for improving rural livelihoods. The project design and implementation should account for the gender-sensitive approach to be successful. A lot of research, studies on implementation, evaluation of existing interventions are required to fully understand the developmental and empowerment implications of ICTs. Reaping the benefits of ICT intervention is a long process but holds the key to empower the women farmers. It's also important to note that the digital divide is not only concerned with technological infrastructure and connectivity. It is critical that ICT initiatives target women and men, as well as the larger family unit and the community to ensure long-term sustainability. An inclusive approach to ICT initiatives will help to generate widespread recognition that it is important for women to be able to use ICTs.

12.3 Green Technologies and Sustainable Agriculture

The concept of a sustainable farming system refers to the capacity of agriculture over time to contribute to overall welfare by providing sufficient food and other goods and services in ways that are economically efficient and profitable, socially responsible, while also improving environmental quality. Organic farming production is typically dominated by small-scale farmers—many of whom are women—and relies on techniques such as crop rotation, composting, and biological pest control (in place of fertilizers and pesticides). The use of genetically modified organisms is also prohibited. Organic farming can thus use locally available inputs and technologies. Demand for organic food has been expanding rapidly worldwide due to growing health and environmental concerns, and the trend is expected to continue (Willer et al. 2019). As women tend to have limited access to chemicals and other inputs associated with industrial agriculture, and thus are more likely to use traditional production systems, with the right assistance they may consider obtaining organic certification to access high-value niche markets, such as those for certified organic products, thus turning a disadvantage into strength (World Bank 2008).

Technological innovation has a potentially important role to play in improving the sustainability of these farming systems through a range of innovations in engineering, information technology, pesticides and biotechnology, reducing the load of known toxins, substituting safer alternatives, protecting ground or surface waters, protecting natural habitats, reducing nutrient loads in soils, reducing gaseous nitrogen loss (Pitkin et al. 1996) or reducing the amount of non-renewable energy used in the cropping cycle. Box 12.3 illustrates the success story of the integrated farming system adopted by women farmers of Kerala.

Sustainable agriculture practices/systems prevalent in India are: Permaculture, Organic farming, Natural farming, System of rice intensification (SRI), Biodynamic agriculture, Conservation agriculture, Integrated farming system (IFS), Agroforestry, Integrated pest management (IPM), Precision farming, Silvipastoral systems, Vertical farming Hydroponics/Aeroponics, Crop-livestock-fisheries farming system, Vermicompost, Drip irrigation/sprinkler, Crop rotation, Intercropping, Cover crops, Mulching, Contour farming, Rainwater harvesting-artificial recharge of groundwater, Floating farming, Plastic mulching, Shade net house, Alternative wet and drying technique (for rice), Saguna rice technique, Farm pond lined with plastic film, Direct seeding of rice, Canopy management Mangrove, and non-mangrove bio-shields (CEEW-FOLU-Sustainable-Agriculture-in-India-2021-20Apr21.pdf).

12.3.1 Constraints for Adoption of Sustainable Farming Practices

The sustainable agriculture practices are not the obvious choice for farmers. Although the sustainable practices have been known to be highly productive and economically competitive, their desirability for farmers may be less due to economic reasons (Horrigan et al. 2002). This may be due to the lead time required to realize the benefits of sustainable farming. The farmers, especially small hold farmers like women will be constrained to make the transition from traditional practices to the sustainable methods.

Many sustainable agriculture practices are complex and may require increased learning for adoption (De Souza Filho 1997). This intellectual cost of understanding the sustainable practices based on farm systems, cropping systems, and others may be heavier on women. Thus, they may not be attracted to changes in farming practices.

Lack of clear information and its dissemination about sustainable agriculture practices is often regarded as a barrier to adoption. Information asymmetry can be due to several sources of information, lack of contextualization, etc. The economic and technical issues, viability of these practices are often not discussed with the farmers. The adoption may be hindered or delayed due to one of these reasons.

12.3.2 Government Initiatives for Sustainable Agriculture in India

National Mission for Sustainable Agriculture (NMSA) has been formulated for enhancing agricultural productivity especially in rain fed areas focusing on integrated farming, water use efficiency, soil health management, and synergizing resource conservation. The aim of promoting sustainable agriculture through a series of adaptation measures focusing on 10 key dimensions encompassing Indian agriculture, namely; "Improved crop seeds, livestock and fish cultures," "Water Use Efficiency," "Pest Management," "Improved Farm Practices," "Nutrient Management," "Agricultural insurance," "Credit support," "Markets," "Access to Information," and "Livelihood diversification." Since 2014–2015, India has invested in NMSA to promote sustainable agriculture. A report on sustainable Agriculture in India 2021 mentions despite several policies of the government, organic farming currently covers only two percent of the country's total net sown area (https://www.ceew.in/publications/sustainable-agriculture-india). This report also sheds light on the very small, merely about 0.8 percent, budgetary allocation to NMSA by the Ministry of Agriculture and Farmers Welfare (MoAFW). Some sustainable agriculture practices receive budgetary support under various Central government programs. Organic farming, integrated agricultural systems, rainwater harvesting, contour farming (terraces), vermicomposting, mulching, precision farming, and IPM are examples of these practices. Organic farming has gotten the most policy attention of them, with some Indian states enacting specific organic farming regulations.

All farming systems, from intensive conventional farming to organic farming, have the potential to be locally sustainable. Whether they are in practice depends on farmers adopting the appropriate technology and management practices in the specific ago-ecological environment within the right policy framework.

12.4 Conclusion

The interplay between gender, technology, economy, and policy in the domain of agriculture is a complex landscape. On the one hand, women face barriers in access to existing technologies, while on the other hand, technological innovation often does not take into account women's needs and economic roles. While there are several policies reinforcing the use of technologies, the socio-economic conditions do not support adoption. Agriculture—encompassing not only raising crops, fish, and animals, but also activities related to food processing and packaging—is a key sector for women's livelihoods around the world. Agricultural technologies can raise women's productivity, providing them with the time to engage in income-generating activities and the capabilities to potentially meet international standards and access higher-value markets.

To benefit from agricultural innovations and technical change, it is important to evaluate how in each context of regional settings (infrastructure, finances, culture, etc.) what corrections can be made to reduce the gender gap. The adoption of new and cutting-edge technologies has been the subject of discussion of many studies across different disciplines. The researchers in agriculture should gain cross disciplinary insights from sociology, economics, political sciences, communication, and public health, to devise suitable recommendations. This should work in parallel with the gender sensitization trainings, collection of sex disaggregated data to help shed light on the various forms of gender inequality in agriculture and inform on gender-sensitive policies which can lower the technology adoption barriers for the women farmers.

References

Achandi EL, Mujawamariya G, Agboh-Noameshie AR, Gebremariam S, Rahalivavololona N, Rodenburg J (2018) Women's access to agricultural technologies in rice production and processing hubs: a comparative analysis of Ethiopia, Madagascar and Tanzania. J Rural Stud 60:188–198

Committee Report (1998) National task force on technological empowerment of women in agriculture. Retrieved January 20, 2022, from http://ncwapps.nic.in/pdfReports/National%20Task%20force%20on%20Technological%20Empowerment.pdf

De Souza Filho HM (1997) The adoption of sustainable agricultural technologies: a case study in the State of Espirito Santo, Brazil. Ashgate Publishing Ltd.

Economic Survey (2017–2018) Ministry of Finance, Government of India. Retrieved January 20, 2022, from https://pib.gov.in/Pressreleaseshare.aspx?PRID=1518099

Gender in Agriculture Source Book (2009) World Bank, FAO, and IFAD. Retrieved January 20 from https://documents1.worldbank.org/curated/en/799571468340869508/pdf/461620PUB0Box3101OFFICIAL0USE0ONLY1.pdf

Gender Mainstreaming in ICT for Agriculture (2012) USAID. Retrieved January 20 from https://pdf.usaid.gov/pdf_docs/pnaec808.pdf

Hansda R (2017) Small-scale farming and gender-friendly agricultural technologies: the interplay between gender, labour, caste, policy and practice. Gend Technol Dev 21(3):189–205

Horrigan L, Lawrence RS, Walker P (2002) How sustainable agriculture can address the environmental and human health harms of industrial agriculture. Environ Health Perspect 110(5):445–456

ICAR (n.d.) Drudgery reducing technologies for women in agriculture. ICAR- Central Institute of Agricultural Engineering, Bhopal. Retrieved January 20, 2022, from https://farmech.dac.gov.in/Women%20Friendaly%20Equipment-With%20photographs.pdf

IFAD (2005) Annu Rep. Retrieved January 20 from https://www.ifad.org/documents/38714170/39624783/ar2005_e.pdf/58a9766e-bf2a-40d3-9a72-73dbc0cbe759

Kale RB, Rohilla PP, Meena MS, Wadkar SK (2015) Information and communication technologies for agricultural knowledge management in India. J Glob Comm 8(1):16–22

Kamala S, Jyothi U, Rani N (2019) ICT availability and usage by farm women in agriculture. Int J Curr Microbiol App Sci 8(11):1620–1629

Lambrou Y, Piana G (2006) Gender: the missing component of the response to climate change. FAO, Rome, pp 1–58

Mathur H (2017) How to encourage the adoption of Agtech innovation in India. Retrieved January 20 from https://agfundernews.com/how-to-encourage-the-adoption-of-agtech-innovation-in-india

Meena J, Seetharaman RN (2004) Cable television viewing behavior of farmers and farm women. IJEE 40:49–55

Murray U, Gebremedhin Z, Brychkova G, Spillane C (2016) Smallholder farmers and climate smart agriculture: technology and labor-productivity constraints amongst women smallholders in Malawi. Gend Technol Dev 20(2):117–148

Oxfam (2018) Gender justice. Oxfam India. Retrieved January 20, 2022, from https://www.oxfamindia.org/women-empowerment-india-farmers

Pitkin PH, Tait EJ, Dawson KP, Cook KA, Fish N (1996) Targeted inputs for a better rural environment: farmers' views of new technology. In: Proceedings Crop Protection in Northern Britain, 1996

Ragasa C (2012) Gender and institutional dimensions of agricultural technology adoption: a review of literature and synthesis of 35 case studies

Rola-Rubzen MF, Paris T, Hawkins J, Sapkota B (2020) Improving gender participation in agricultural technology adoption in Asia: from rhetoric to practical action. Appl Econ Perspect Policy 42(1):113–125

Singh PK, Tiwari R, Dutt T, Joshi P (2018) Adoption of ICT among the farmers of different agro climatic zones of Uttar Pradesh. Int J Curr Microbiol Appl Sci (7):5250–5255

Sinyolo S (2020) Technology adoption and household food security among rural households in South Africa: the role of improved maize varieties. Technol Soc 60:101214

Van Eerdewijk A, Danielsen K (2015) Gender matters in farm power. KIT, CIMMYT and CGIAR, Amsterdam

Vemireddy V, Pingali PL (2021) Seasonal time trade-offs and nutrition outcomes for women in agriculture: Evidence from rural India. Food Policy:102074

Vijaya LV, Deepika J (2020) Awareness on drudgery reducing farm technologies for gender equity. Asian J Res Agric For. Retrieved January 20 from https://www.journalajraf.com/index.php/AJRAF/article/view/30110/56501

Willer H, Lernoud J, Huber B, Sahota A (2019) The World of Organic Agriculture, Statistics and Emerging Trends 2019 at BIOFACH 2019

World Bank, Food and Agriculture Organization and International Fund for Agricultural Development (2008) Gender in agriculture sourcebook. The World Bank

Correction to: Extent of Participation of Farm Women in Decision Making Regarding Agricultural Activities

Anuradha Ranjan Kumari, Barun, S. K. Mandal, and R. K. Mandal

Correction to:
Chapter 10 in: S. Munshi, M. Singh (eds.), *Women Farmers: Unheard Being Heard*, Sustainable Agriculture and Food Security, https://doi.org/10.1007/978-981-19-6978-2_10

The original version of this chapter was inadvertently published with incorrect affiliation of the authors, The affiliation has been updated with this correction. The correction affiliation should read as

Anuradha Ranjan Kumari, Barun, S. K. Mandal and R. K. Mandal
Krishi Vigyan Kendra (DrRPCAU) Bhagwanpur Hat, Siwan, Bihar, India

The updated version of this chapter can be found at
https://doi.org/10.1007/978-981-19-6978-2_10

S. Munshi, M. Singh (eds.), *Women Farmers: Unheard Being Heard*, Sustainable Agriculture and Food Security, https://doi.org/10.1007/978-981-19-6978-2_13

The manufacturer's authorised representative in the EU is Springer Nature Customer Service Centre GmbH, Europaplatz 3, 69115 Heidelberg, Germany. If you have any concerns regarding our products, please contact ProductSafety@springernature.com

Printed and bound by CPI Group (UK) Ltd, Croydon, CR0 4YY
30/09/2026
02225265-0017